Beginner's Guide to
Technical Illustration

Beginner's Guides are also available on the following subjects:

Audio
Building Construction
Cameras
Central Heating
Colour Television
Computers
Digital Electronics
Domestic Plumbing
Electric Wiring
Electronics
Gemmology
Home Energy Saving
Integrated Circuits
Photography
Processing and Printing
Radio
Super-8 Film Making
Tape Recording
Television
Transistors
Video
Woodturning
Woodworking

Beginner's Guide to
Technical Illustration

Clive Holmes

Newnes Technical Books

Newnes Technical Books
is an imprint of the Butterworth Group
which has principal offices in
London, Sydney, Toronto, Wellington, Durban and Boston

First published 1982

© **Butterworth & Co (Publishers) Ltd, 1982**

All rights reserved. No part of this publication may be reproduced or transmitted in any form or by any means, including photocopying and recording, without the written permission of the copyright holder, application for which should be addressed to the Publishers. Such written permission must also be obtained before any part of this publication is stored in a retrieval system of any nature.

This book is sold subject to the Standard Conditions of Sale of Net Books and may not be re-sold in the UK below the net price given by the Publishers in their current price list.

British Library Cataloguing in Publication Data

Holmes, Clive
   Beginner's guide to technical illustration
   1. Mechanical drawing
   I. Title
   604.2'4076   T354

ISBN 0-408-00582-3

Typeset by Butterworths Litho Preparation Department
Printed in England by Mansell (Bookbinders) Ltd., Witham, Essex

Preface

In writing this book my aim has been to provide, in simple language and with the aid of clear illustrations, a means by which the layman can understand some of the complexities that beset the technical illustrator. The reader can work through in simple stages, using the text and diagrams as a guide from one exercise to another, so that eventually he or she is able to produce simple technical illustrations to the standard set by the assessors of the new Technician Education Council (TEC III).

Simple technical illustration cannot be learned by merely reading a book on the subject. Like drawing, like welding, like cooking, it is a practical subject, a 'doing' subject and it must be pursued in a practical manner. To learn how to produce technical illustrations, therefore, the student must get on the drawing board and illustrate. There are books on the subject, if you can find them, but few have been produced recently and most are written in such a way as to make very heavy going of the subject. This book, by contrast, attempts to make the subject as understandable as possible for the reader who has no prior knowledge. This philosophy has been combined with stage by stage diagrams of 'How to do it' followed by a series of exercises to get home the messages learned. The aim is somewhere between a text book and a work book on simple technical illustration. Since the City and Guilds of London Institute is in the process of phasing out as the Examiners in this subject, then ultimately a course of Technician Education Council Year IV or the equivalent level should be the aim of all those wishing to become industrial illustrators. This book will cover much of the foundation work for those whose aim is a career in the subject, whilst also helping all other students of the subject, vocational and non-vocational alike, and should prove of value to students of engineering, draughting and graphic design of CSE, GCE & TEC work levels.

The subject of 'Technical Graphics' or 'Industrial Art' is far too full a subject to squeeze between the covers of a book in the 'Beginners Guide' series. Indeed it would take a more elaborate and expensive work to cover the subject properly. Through necessity it would include colour and not contain itself merely to black and white. This book then is, as the title says, a 'beginner's guide'; it is not the bible on the subject but is based on much varied industrial and lecturing experience in the UK and the USA at both junior and senior level. It is meant as an introduction to the practicalities of the core of these subjects, that core being the illustration of technical matter.

Many of the illustrations in this book have been produced by my students, and no illustration is of a subject more complex than the level of work required by the TEC Years II/III assessment.

Finally it should be noted that, contrary to the belief of many people not engaged in this country's industries, many companies still function in feet and inches, whilst others, mainly the larger ones, have gone metric. It is necessary therefore that future technical illustrators be familiar with both sets of dimensioning. Whilst most younger people will be familiar with metric dimensioning at school, they are probably not so proficient with the Imperial system. The dimensioning in this book uses both systems.

C.H.

Contents

1 What is technical illustration?   1

2 Orthographic information   8

3 Objects in three dimensions   30

4 A formula for technical illustration   45

5 An introduction to perspective   75

6 Basic techniques   84

7 Let's illustrate   109

8 Free-hand drawing and sketching   139

9 Tools of the trade   150

    Index   165

1 What is technical illustration?

'One illustration is worth a thousand words'. 'A thousand words is worth one illustration'. The age-old arguments between the artist and the writer still resound in many technical publications departments. On one hand are those who believe it is better to communicate using illustrations, and on the other hand are those who believe that it can better be done with the written word. The ideal must surely be a combination of both!

The monotonous manner in which text book information was presented in the past has largely been replaced by a more serious means of communication with the 'thousand words' being combined with the 'one illustration'. Today much depends on the operation and maintenance of machinery of one kind or another; from the simplest lawn mower to the most complicated missile system. In every area of modern industry more use is being made of the person who can combine artistic ability and mechanical aptitude to communicate information by visual means.

The technical artist is employed in the production of artwork for sales and publicity purposes, maintenance, spare parts and operation manuals in addition to general educational literature. Although much artwork is produced in a two- dimensional form, often called 'flatties', most illustrations are of a three-dimensional nature; the form easiest for the human eye to accept is perspective. Sometimes the illustration takes the form of a 'cut-away', where part of the object is removed to show the inner workings of an assembly. Figure 1.1 shows a detail of a micrometer, the outside casing of the barrel having been removed in order to show its inner workings. In this instance the cut-away edge has been treated as if broken so as not to confuse the viewer with the mechanisms inside the barrel. Figure 1.2 shows the opposite type of illustration, the exploded type, when each part of the assembly is exploded along its central axis normally in order of disassembly; that is to say what went on last comes off first.

Although some technical illustrations are produced in colour the vast majority are line drawings, some of which are shaded to distinguish different parts, textures and materials, although in the main shading is kept to the minimum to keep the cost down. It was often said of the older illustrations that the vast amount of shading used by the artist to describe the subject was a means of

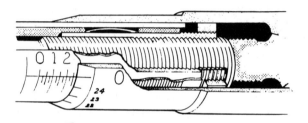

Figure 1.1. Cut-away illustration showing detailed workings of a micrometer barrel

covering up 'a multitude of sin'. This was true in some cases, as the perspective and formation of some parts was so poor that the only way the artist could make it look right was by putting on lots of shading and highlights. Nowadays most industrial illustrations are of a high standard and communicate well the size, shape and form of the subject to the viewer without a lot of shading. Often the illustrations are required before the product has actually been completed; indeed spares and maintenance manuals are often part of the sales package and must be ready at the same time as the product itself. Occasionally illustrations are required before the product even starts production, as part of the advance marketing information. Such illustrations are often very elaborate as much advance data can be gained depending on the way in which they are used. A simple black and white illustration

WHAT IS TECHNICAL ILLUSTRATION?

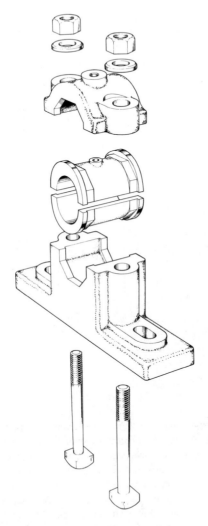

Figure 1.2. An exploded illustration

produced for such a purpose is shown in Figure 1.3. Its finish is in line with tones to indicate the finish of the control units together with the tinted translucent body and its overall appearance, and was required in order to establish the value of this particular design's aesthetic qualities.

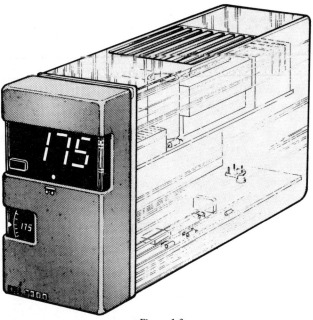

Figure 1.3

One of the first and last commandments that the junior illustrator learns is 'Thou shalt not shade'. This is true to a great extent, but there are always exceptions to the rule. We are all familiar with the illustrations produced in motor and aircraft magazines. To work on this kind of subject is the ambition of many a technical illustrator, but such positions are open only to the fortunate few. The work produced for these magazines compared with the majority of technical publications work is very glamorous. It must appeal to the eye as well as be technically correct. In order to make the car or the aircraft look attractive the illustrator frequently adds shading that would not be allowed if he were working for a different market. Whilst understanding that generally shading is kept to the minimum there may be a time when it becomes important to know how it is done. Thus some understanding of shading is necessary for all illustrators.

Background information

The reason why people are more conscious today than in the past of the business of 'visual communication' must obviously have something to do with the industrial age in which we live. There is also the fact that the vast majority of people in the western world can read and write, with a large percentage of the natives of the Western European countries able to communicate in more than one language. We all have access to newspapers, magazines, books, radio and television and as we are bombarded with facts and information from all sides we are able to communicate and to understand a vast amount of subject matter. The picture however was very different if we look back into history even a short way, say about the time of the First World War. The population of the same part of the world was nowhere near being so 'technically orientated', the vast majority of people made their living from the land and the technological know-how assimilated by the masses of today was largely unknown except for the minority who could afford a good education. Although it is fairly new, we do have women engineers, pilots and drawing office personnel. In 1914, however, very few women could read a newspaper let alone read an engineering drawing. It was against this background and with the contribution of many factors that the Great War broke out. On both sides the men went to war and on both sides the women went to work, some on the land and others in the munitions factories. Most of these women worked as non-skilled labour on repetitive production lines assembling engineering parts. Since few had any idea of orthographic drawing, simple illustrations were used to indicate

6   WHAT IS TECHNICAL ILLUSTRATION?

what went where. A simple picture was produced probably by a draughtsman to indicate the sequence of assembly of certain parts. This then by-passed the need for an understanding of plan, front and end elevations. It became apparent that even the illiterate could understand a picture. I suppose in an instance such as this an illustration was truly worth more than a thousand

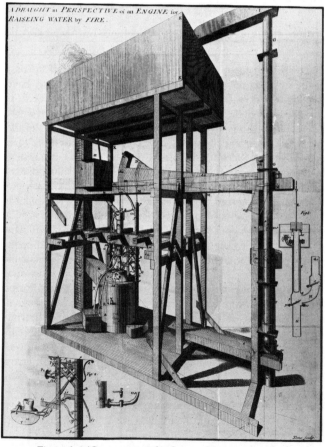

Figure 1.4 (Crown copyright. Science Museum, London)

words. This system, then, was the forerunner of the product illustration used extensively during the fifties and sixties in the USA to inform semi-skilled assemblers how to put together certain pieces of equipment. It is still used in certain industries today where other methods of communication may only confuse the issue. We can of course go deeper into history and find more evidence. Leonardo da Vinci and Archimedes are two names that immediately spring to mind when considering technology of yesteryear. Both of these people had a need to communicate their ideas, theories and designs. They wrote down their findings but they also drew and on many occasions they annotated their drawings so that the viewer could understand their findings in two ways; by the written word and by visual image. In short they then became visual communicators of technical information.

There are many historical instances of people informing others by means of a picture of how a process works or how a machine functions, but although a history of visual communication of technical information would undoubtedly be a very interesting subject it is not the purpose of this book to follow this through. Nevertheless it is of interest to take a look at Figure 1.4. This is an English illustration of the Newcomen waterworks pumping plant and is dated 1720. The perspective is exaggerated and no indication of the scale is given but since it has letter keys on the drawing they must refer to written specifications in the text. The artist has included a few detail drawings to clarify certain of the more complicated areas of his design. Note also the amount of shading he has employed: it must have taken almost as much time to shade as to complete the initial illustration. This is just one of many examples of types and styles of illustrations of yesteryear.

2 Orthographic information

For the average person who wishes to understand technical illustration, there are two main important stages in his or her development. The first is learning to read and write in the early days at school; obviously without this ability the learning potential is very limited. The second is learning to read an engineering drawing, for without this knowledge two-dimensional information cannot be understood, neither can it be transmitted into three dimensions. The child at school, before learning to read and write, has only a limited means of communication. After learning to read and write a whole world of information becomes available. When orthographic information can be understood by the would-be technical illustrator, then vast amounts of technical information can be absorbed, digested and distributed in a manner that is easily acceptable to the human eye. Until this ability is attained there is obviously a limit to the amount that can be achieved as a technical illustrator. The point being made is that the first stage in learning to illustrate is learning to read orthographic information.

The basic views

Orthographic drawing is a method by which an object can be represented in two dimensions. Basically three views are needed to show an object clearly, these being the plan, the front elevation and the end elevation. For more complex objects it is often necessary to show additional views, sectional views and auxiliary views. Since we are starting at the beginning of the subject and dealing only with simple objects in the first instance,

ORTHOGRAPHIC INFORMATION

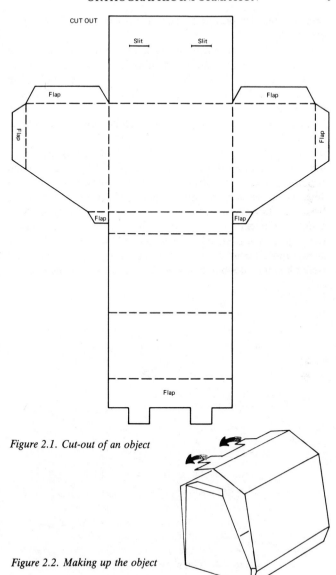

Figure 2.1. Cut-out of an object

Figure 2.2. Making up the object

we will consider only the basic views of *plan, end* and *front elevations*.

Figure 2.1 is a cut-out from which a three-dimensional object can easily be made to clarify orthographic projection. In order to avoid damaging this book trace off the cut-out and 'push through' on to a light card. To 'push through' refer to Chapter 7 for details.

To use this object, first of all carefully cut out round the outline. Next slit the two slots marked 'slit' and then fold backwards over a rule (to ensure a straight fold) along the dotted lines. Then fold all the faces of the object in (Figure 2.2) and tuck into the slots the tabs marked 'X'. You now have an object with which we can examine orthographic projection and the basic elevations of plan, front elevation and end elevation. Now stand the object on its base as in Figure 2.3. If we look down on the object in the direction of arrow 'X' we can see the view shown above arrow 'X': this is the plan of the object. Likewise if we look in the direction of arrow 'Y', we can see the view shown

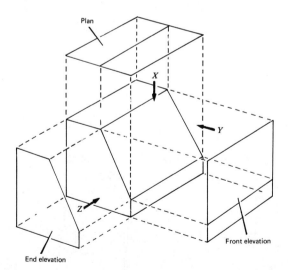

Figure 2.3. How we see the different views

ORTHOGRAPHIC INFORMATION

behind arrow 'Y' which is the front elevation. Finally if we look in direction of arrow 'Z', we can see the view shown behind arrow 'Z' which is the end elevation.

Take the three-dimensional object now and place it with its base downward on a flat surface (Figure 2.4). The viewer should then look down directly onto the top of it to see the plan view (A). Now rotate the object forwards slowly (B) until the front view becomes apparent (C). Next, rotate it slowly through position (D) to (E). This will give the left end elevation. Then finally after placing it back in position (C) rotate it slowly through position (F) to (G) to give the right end elevation.

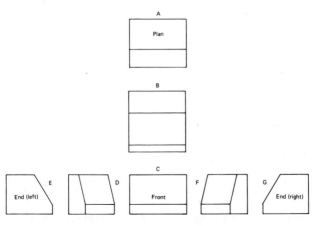

Figure 2.4. How we see the different views of our object

Now go back to the beginning of this chapter and read the text under the heading 'The basic views' again a few more times in conjunction with Figures 2.1, 2.2, 2.3 and 2.4 until you understand each view in relation to the others. When not in use for clarification of views, the three-dimensional object should be folded flat and paper-clipped flat to the page that you used last. Used this way it will be kept flat and can be used as a book marker.

First and third angle projection

Let us now put aside for a while the three-dimensional object and look at a different shape, the simple right angle bracket shown in Figure 2.5. If we were to lay this on its base as shown in the sketch, the plan, front elevation and end elevation would be as shown in Figure 2.6. You will note that the change of plane lines marked 'X' go together via all the views but only as hidden detail (indicated by dotted lines) in the other views, while the hole centre and its centre axis is shown by the chain dotted line (long and short dots). Observe that the radiused end to the bracket base, like the hole, has the same centre and can only be seen in the plan view.

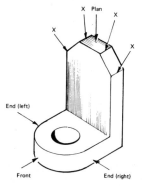

Figure 2.5. A simple right angle bracket

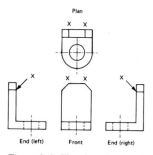

Figure 2.6. The plan, front and end elevations

Having dealt with the different elevations of the object we now have to learn how to lay out the various views in a standard, easily read, logical pattern or convention. There are two methods by which this can be done: one is called 'First Angle Projection' and the other is called 'Third Angle Projection', the former sometimes being referred to as 'British Projection', and the latter as 'American Projection'. Until the Second World War, First Angle or British Projection was used almost entirely in the UK, but as our engineers and designers liaised more and

ORTHOGRAPHIC INFORMATION 13

more with the engineers of the USA, their system of Third Angle Projection became adopted and is now more commonly used in this country. The British Standards Institution therefore permits the use of both methods.

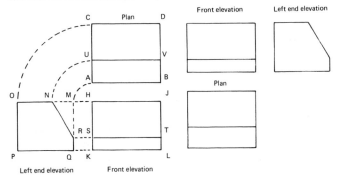

Figure 2.7. Third angle projection   *Figure 2.8. First angle projection*

If we now consider our three-dimensional object, and lay out a plan, front and end elevation of it as shown in Figure 2.7, then this is Third Angle Projection. If we lay out the same object in the way shown in Figure 2.8 then it conforms to First Angle Projection (end elevations being projected to the opposite side of the front elevation to which they actually are).

Third angle projection

As Third Angle Projection is the simplest for the layman to understand let us consider it before we deal with First Angle Projection. Referring to Figure 2.7 it is clear that the plan is drawn above the front elevation and the end elevation is swung round and projected to the side of the front elevation. To produce a Third Angle Projected drawing of this object, first of all draw the horizontal line A–B 3in (76mm) long. From points A and B draw verticals A–C and B–D each being 2in (51mm) high. Link points C and D by a horizontal line C–D, completing the outline of the plan. Draw two lines parallel to A–B, the first 1 in below A–B, the second 3 in (76 mm) below A–B. Produce C–A and D–B to cut these two lines at H–K and J–L respectively. We now have an outline shape of the front elevation which conveys to the reader the length and the height. Now with this information and the three-dimensional object we can proceed to the production of the end elevation.

With the compass-point on point H, open the compass out until the lead touches point A and swing an arc of a circle from A about H to cut a line produced from J–H; this point we will call M. Having produced point M, then in similar manner produce point O. Produce L–K and cut it at points P and Q by dropping verticals from O and M. Next, measure up the vertical Q–M ½ in (12.7 mm) from Q to give us the height of the vertical face in the end elevation; call this point R. Likewise the width of the horizontal top of the object will be 1½ in (38 mm). In the end elevation this can be shown by measuring 1½ in (38 mm) along O–M and marking off point N. By linking N–R we now have the chamfer as seen in the end elevation. In the front elevation produce a line from point R to cut H–K and J–L at points S and T. This will complete the bottom of the slope in this view. If point N in the end elevation is swung about point H to cut C–A at U, then by drawing a line to cut D–B at V the whole orthographic drawing is complete. To sum up, we can say that in Third Angle Projection, if the plan of the object is drawn first, then we must ask ourselves – 'If we look in the direction of arrow 'Y' (Figure 2.3) what do we see?' We see the front elevation of the object, below the plan, not above it. Therefore we draw it below the plan. If we now look into the end of the front elevation (arrow 'Z' Figure 2.3) what do we see? We see the end elevation. Therefore we draw the end elevation alongside the front elevation.

First angle projection

As we have worked through, step-by-step, the production of a Third Angle Projected drawing of our three-dimensional object, there would seem little need to repeat so much to produce the First Angle Projected view. The stages followed would be similar to those followed in projecting the Third Angle drawing but all views in First Angle are laid out as indicated in Figure 2.8. The plan goes beneath the front elevation. Let us commence by producing the front elevation, then ask ourselves what we see if we look down on the object in direction of arrow 'X' (Figure 2.3)? The answer is the plan. The plan is now drawn *below* the front elevation and the end elevation (view on arrow 'Z' Figure 2.3) is projected to the opposite side of the object to the side seen (Figure 2.8). Where Third Angle Projection can be called the logical layout then First Angle Projection *could* be called the illogical layout.

Orthographic exercises

On pages 16–18 are a total of ten simple shapes. All are missing certain lines, centre lines and hidden details, or have incomplete three-dimensional sketches. Using a medium grade of pencil to obtain a bold line, sketch in, free-hand, the missing information. You should complete a plan, a front elevation and an end elevation (right side) together with a sketch of each object. Don't bother to use drawing instruments yet. Again, if you do not wish to damage or mark this book, trace off each page and complete the work on the traced images. *On completion* check your answers at the end of this chapter but remember you won't learn if you cheat.

Orthographic exercises

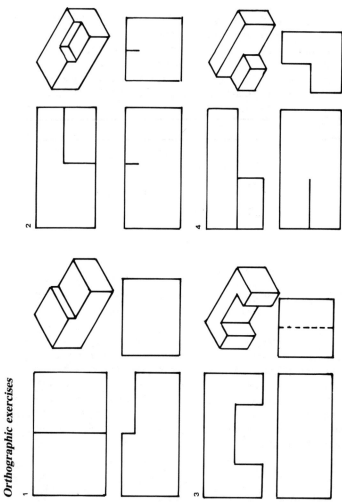

ORTHOGRAPHIC INFORMATION 17

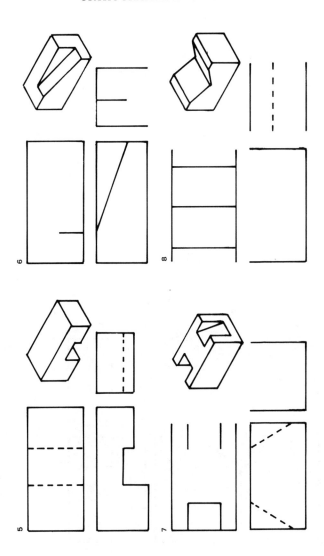

ORTHOGRAPHIC INFORMATION

Orthographic exercises (contd)

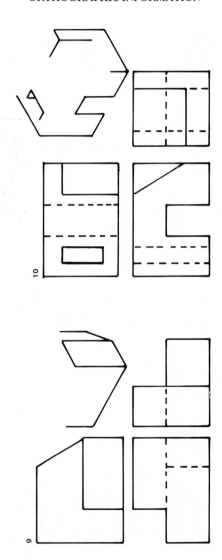

Draughting exercises in first and third angle projection

On completion of the ten simple shape exercises then elementary draughting can be studied. The following pages contain three exercises laid out in Third Angle Projection. These should be worked through from No. 1 to No. 3. You will notice that these exercises contain only two views: in each case one of the three views is missing. Starting with No. 1 re-draw the two given views and complete the third view. When you have completed three views of the object, laid out in Third Angle Projection, then re-draw the object laying it out in First Angle Projection. When you have produced the layouts in first and third angle projection of object No. 1 then move on to No. 2 and finally No. 3.

Each view must be projected one from the other as indicated in Figure 2.7. Widths of objects are swung round from the plan view using the compass until intersecting with the horizontals extended from the front elevation and so on until all three views are complete as described in the text on Third Angle Projection. Each vertical line must be *VERTICAL* and each horizontal line must be *HORIZONTAL* and of course be at right angles to each other. All dimensioning is to be included. The types of line to be used *must* conform with the details of the types of line specified in British Standard 308 (Figure 2.11, pp. 28, 29).

ORTHOGRAPHIC INFORMATION

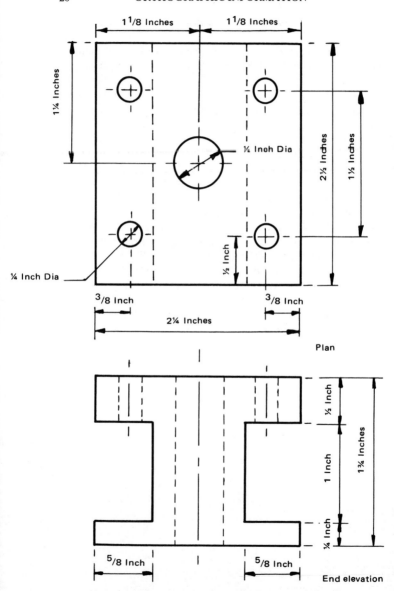

Exercise 1. Produce views in first and third angle projection

ORTHOGRAPHIC INFORMATION

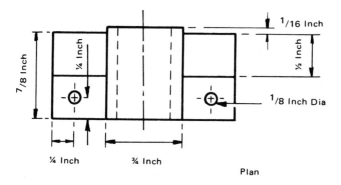

Plan

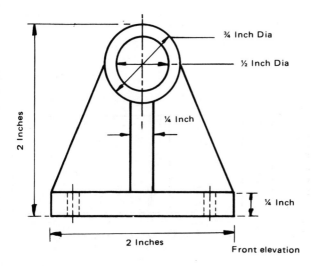

Front elevation

Exercise 2. Produce views in first and third angle projection

ORTHOGRAPHIC INFORMATION

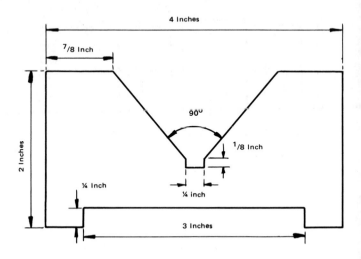

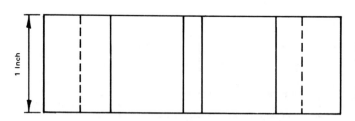

Exercise 3. Produce views in first and third angle projection

Draughting tips

If you are using a Tee-square make sure it is firmly held against the edge of the drawing board (Figure 2.9) to ensure accuracy. Pencils can be of wood or the more modern clutch on fine line lead holders (mechanical) but if either of the two former types

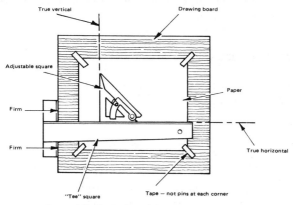

Figure 2.9. Basic draughting equipment in use

are used, then keep them sharp by using a sharpener or a sandpaper block. The centre point of your compass should have a shoulder (Figure 2.10) to prevent the user from digging holes in the drawing.

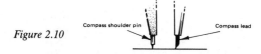

Figure 2.10

If, however, holes are 'dug' then try a small piece of masking tape over the damaged area. This will reinforce the surface and accuracy will be maintained. Usually a shoulder point is found on the opposite end of the long conical point used normally for dividers. Generally the shoulder point should extend about half its length beyond the lead end. When pencilling, it is easiest to hold the pencil at 50°–60° to the paper in a plane parallel to the edge surface of the Tee-square or adjustable square. This will become apparent as experience is gained. Don't grip the pencil like a hatchet; hold it firmly but flexibly. Initially all line work should be light using, say, 4H or 3H lead, but completing with a softer lead. The finished drawing should be in a firm, compressed 'unfluffed edge' black line, from which a clean and clear reproduction could be made.

Conversion tables

The following figures are given to the nearest $\frac{1}{10}$ mm and the nearest $\frac{1}{1000}$ inch.

Inches to millimetres

1/16	–	1.6
1/8	–	3.2
3/16	–	4.8
1/4	–	6.4
5/16	–	7.9
3/8	–	9.5
7/16	–	11.1
1/2	–	12.7
9/16	–	14.3
5/8	–	15.9
11/16	–	17.5
3/4	–	19.1
13/16	–	20.6
7/8	–	22.2
15/16	–	23.8
1	–	25.4
2	–	50.8
3	–	76.2
4	–	101.6
5	–	127.0
6	–	152.4
7	–	177.8
8	–	203.2
9	–	228.6
10	–	254.0
11	–	279.4
12	–	304.8

Millimetres to inches

1	–	0.039
2	–	0.079
3	–	0.118
4	–	0.157
5	–	0.197
6	–	0.236
7	–	0.276
8	–	0.315
9	–	0.354
10	–	0.394
11	–	0.433
12	–	0.472
13	–	0.512
14	–	0.551
15	–	0.591
16	–	0.630
17	–	0.669
18	–	0.709
19	–	0.748
20	–	0.787
21	–	0.827
22	–	0.866
23	–	0.906
24	–	0.945
25	–	0.984
26	–	1.024

ORTHOGRAPHIC INFORMATION

Answers to orthographic exercises

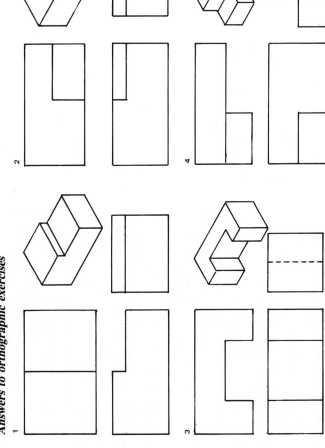

ORTHOGRAPHIC INFORMATION

Answers to orthographic exercises (contd)

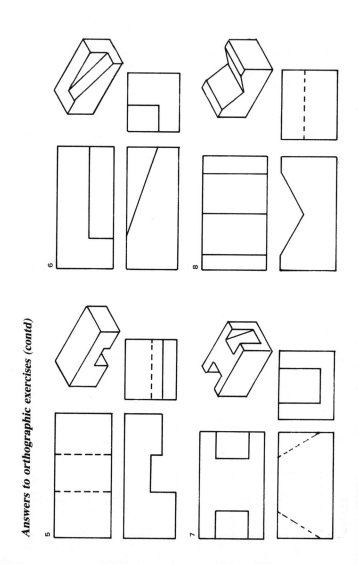

ORTHOGRAPHIC INFORMATION

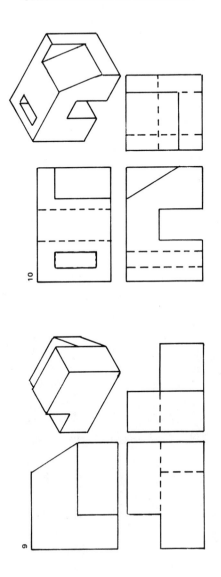

ORTHOGRAPHIC INFORMATION

Type of Line		Example	Application
Continuous (thick)	A	————	Visible lines
Continuous (thin)	B	————	Dimension lines / Projection or extension lines / Hatching or sectioning
Short Dashes (thin)	C	- - - - -	Hidden details
Long Chain (thin)	D	—·—·—·—	Centre lines / Pitch circles
Long Chain (thick)	E	—·—·—·—	Cutting or viewing planes
Short Chain (thin)	F	—·—·—·—	Developed or false views / Alternative position of movable part
Continuous wavy (thick)	G	～～～	Short break lines
Ruled line and short zig-zags	H	—⋀—⋀—	Long break lines

ORTHOGRAPHIC INFORMATION

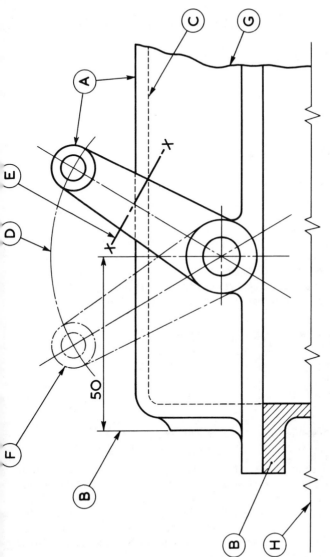

Figure 2.11. Types of lines which should be used for general engineering drawings
(courtesy British Standards Institution)

3 Objects in three dimensions

From circle to ellipse

If we look down directly onto the centre point of a square which contains a circle then we can see it as shown in Figure 3.1. If the eye is now moved down to left or right of the square (Figure 3.2) the square appears to be foreshortened in varying degrees and so does the circle inside it, depending on how far away from the eye the square is set. The lower down we come the more foreshortened the object appears. For instance, if the object is viewed from point A it appears as a plan view (Figure 3.1) but if viewed from point B then it appears more like Figure 3.3.

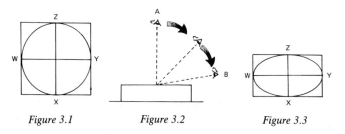

Figure 3.1   *Figure 3.2*   *Figure 3.3*

Further, if the square with the circle inside is held in a vertical position and is viewed from a position directly in line with its centre point then it appears as in Figure 3.1. If however the square is turned to left or right, so that its face is angled toward the viewer instead of its face being at right angles to the viewer, then it will again appear as a foreshortened square with a foreshortened circle inside it (Figure 3.4).

OBJECTS IN THREE DIMENSIONS

We could say, then, that there is only one position from which we could view a circle and still see a circle; that position is at right angles to its face and having our eyes exactly in line with its centre point. In any other position we would see a foreshortened circle, or, in other words, an ellipse.

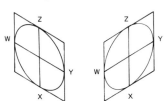

Figure 3.4

In Figure 3.1 you will observe that the horizontal W–Y and the vertical Z–X are the same length and are at right angles to each other because the shape of the object is true. In Figure 3.3 the object is foreshortened so the horizontal W–Y stays the same length, whereas the vertical Z–X becomes foreshortened. It should be noted now that at whatever angle to the viewer our object is turned the angle of intersection of these two lines is always 90°, a right angle. Further, the larger of the two, in this case W–Y, is called the major axis; and the smaller of the two, in this case Z–X, is called the minor axis (Figure 3.3).

How we see the ellipse

Suppose we consider a simple cylinder. It has a circular top and bottom both of equal size because the sides of the cylinder are not tapered but parallel to each other. To produce an illustration of this object is fairly straightforward: all that is needed are two ellipses, one at the top and one at the bottom, joined by two parallel lines (Figure 3.5).

If however we consider illustrating a complex piece of machinery, then it is not so straightforward. The illustration could contain gears, shafts, bearings etc., all represented by different sized ellipses. It is necessary, therefore, to be able to construct these ellipses easily and accurately. There are a number of different methods which can be used to construct an ellipse. But since we need not concern ourselves with a lot of unnecessary geometry only two methods are dealt with. Of these, the most useful for our purpose is the Trammel method.

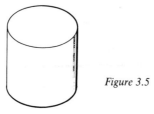

Figure 3.5

Methods of constructing the ellipse

Auxiliary circle method

Having drawn the major and minor axes at right angles to each other, draw a circle equal in diameter to the major axis. Then draw a second circle equal in diameter to the minor axis. Draw lines right across the larger circle; where these lines cut the small circle draw lines parallel to the major and minor axes to produce a series of points 'A'. Then by linking all these points by a curve the desired ellipse can be produced (Figure 3.6).

Trammel method

Draw a major and a minor axis at right angles to each other. Then measure off on a strip of paper (having a straight edge and known as the trammel) the distance O–D. Now place the edge of the trammel along line A–C with point D touching point A, then mark on the trammel point O, calling it on the trammel point X. To plot the points now move the trammel round so that point O on the trammel slides up line A–C. Point X on the trammel slides along line B–D and point D on the trammel is used to plot

OBJECTS IN THREE DIMENSIONS

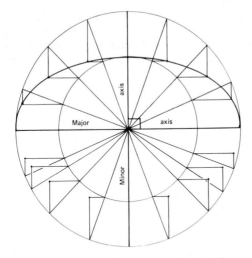

Figure 3.6. The auxiliary circle method of ellipse construction

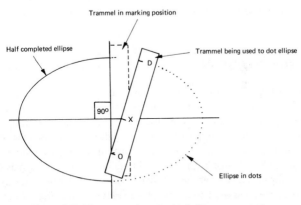

Figure 3.7. The trammel method of ellipse construction

a series of points, and the trammel is moved round in the same direction until the whole ellipse is complete. All that remains is to link the plotted points with a smooth curve (Figure 3.7).

Obviously the closer together the points are placed then the more accurate the finished ellipse will be; and since at this stage we are *not* aiming for quantity but quality, make sure that points are accurately plotted and connected with a smooth, clean curve.

Examples

Produce six ellipses, three by the auxiliary circle method and three by the trammel method, to the following dimensions.

Major axis	Minor axis
6 in	4½ in
203 mm	76 mm
5 in	4¾ in

Be critical of your work and after each exercise stop and look at what you have produced. Decide how you might have improved your work. Do this all the way through; aim to be an accurate illustrator and speed will come with experience.

Ellipses in proportion

Supposing we now wish to produce an ellipse in the same plane and in proportion to an ellipse which we have already drawn. Let us suppose also that the dimensions of the first ellipse are as in Figure 3.8. Now we wish to draw an ellipse of major axis 3 in (76 mm) in the same plane and in proportion to the ellipse in Figure 3.8. We don't know the dimension of the minor axis of this ellipse, but we can find it by drawing a line from major axis to minor axis in the first ellipse A–B (Figure 3.8). Then produce lines parallel to it from the major axis of the smaller ellipse, C–D^1 and C–D^2 (Figure 3.9). Therefore D^1–D^2 is the minor axis and the second ellipse can be constructed using both points C and both points D. Thus the second ellipse, in spite of being a different size, is in proportion to the first ellipse.

OBJECTS IN THREE DIMENSIONS

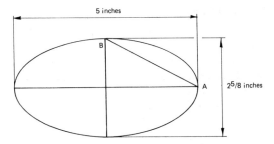

Figure 3.8

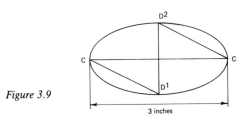

Figure 3.9

If we discount perspective then we can safely say that ellipses in the same plane must be in the same proportion.

Now construct an ellipse of major axis 7 in (177.5 mm) and minor axis 3 in (76 mm). Next construct a major axis of 4¼ in and using the described method work out the minor axis and construct the ellipse. Follow this by completing the exercises at 'A' and 'B'.

Exercises

Produce ellipses in proportion:

Ellipses 'A'

	Major axis	Minor axis
(1)	178 mm	76 mm
(2)	6 in	4½ in
(3)	102 mm	75 mm

Ellipses 'B'

	Major axis	Minor axis
(1)	107 mm	
(2)	5¼ in	
(3)		62 mm

36   OBJECTS IN THREE DIMENSIONS

Isometric projection

Isometric projection is one of the methods by which a three-dimensional illustration can be produced from an orthographic drawing. It is really a means by which a draughtsman, with his ability to calculate and portray an object methodically and precisely, can represent the length, width and height of an object, regardless of whether or not he has any artistic ability. It could be described as a compromise between an orthographic drawing on one hand and an illustration on the other hand.

The advantage of isometric projection over perspective is that it is less time-consuming to modify and up-date illustrations produced in this manner, as every line is either vertical or at an angle of 30° and each ellipse is 35°. Thus certain companies who are continually up-dating their products find it is more economical to use isometric illustrations in technical publications work than the more involved perspective illustrations.

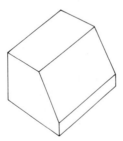
Figure 3.10. Perspective

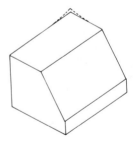
Figure 3.11. Isometric

The prefix 'iso' derives from a Greek word meaning 'equal' – in this case, equal length (of line) and equal angle. This method has, however, the disadvantage that the top corner of an object portrayed this way appears to be lifting up off the surface of the paper, as the eye is used to seeing things in perspective. Compare the difference between the object we dealt with in Figure 2.1 drawn in perspective (Figure 3.10) and in isometric (Figure 3.11). The larger the isometric drawing is, then the more apparent this 'lifting up' becomes.

OBJECTS IN THREE DIMENSIONS

Pure isometric

Let us suppose that we are required to produce an isometric drawing of a 3 in (76 mm) sided cube. Now we all know that the height, length and width of a cube are equal. If we lay out an orthographic drawing of the cube in Third Angle Projection we will see it as shown in Figure 3.12. The 3 in (76 mm) dimension shown in the plan shows that the object is 3 in (76 mm) deep. The 3 in (76 mm) dimension shown on the front elevation shows that the object is 3 in (76 mm) wide and the 3 in (76 mm) dimension in the end elevation shows that the object is 3 in (76 mm) high.

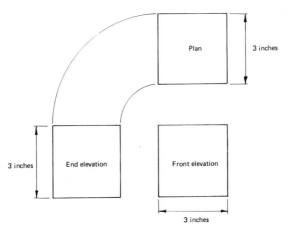

Figure 3.12. The cube in orthographic projection

To start our projection of the cube (Figure 3.13), first draw a horizontal line A–B, say 5 in (127 mm) long with its mid-point C. From point C complete the vertical C–D about 5 in (127 mm) long. From point C mark point E 3 in (76 mm) along C–D. Next, from point C, draw a line C–F about 4 in (102 mm) long at an angle of 30° to the horizontal. Draw a line E–G and drop a vertical line L–M. In the same way complete the left hand side with C–H and J–E with the vertical N–O. Then draw N–K parallel to E–L and K–L parallel to E–N, to complete the cube.

We could now say that face N, E, C, O represents the end elevation from Figure 3.12. E, L, M, C represents the front elevation and N, E, L, K represent the plan. Thus we have a

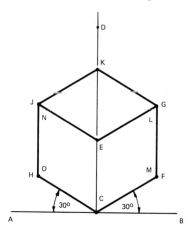

Figure 3.13. The cube in pure isometric projection

three-dimensional drawing of the cube from the information given in the orthographic two-dimensional drawing.

Foreshortened isometric

As was pointed out early in this chapter, isometric projection is a compromise looking wrong because it is not in perspective. A simple form of foreshortening has been worked out which does help to reduce the exaggerated appearance of pure isometric (or plain isometric as it is otherwise called). All vertical dimensions are in standard units of inches or millimetres etc., but all dimensions measured along all lines drawn at 30° to the horizontal, both left and right are foreshortened by using a foreshortening scale. In Figure 3.14 two lines are drawn at angles to the horizontal, the first at 30° and the second at 45°. Now supposing we require a dimension of 3 in (76 mm) foreshortened, then we measure 3 in (76 mm) along line A–C and then project it down by dropping a vertical until it cuts line A–B at point D. The length A–D is the foreshortened dimension we require and this can be transferred with the dividers to where it is required in our illustration. Obviously if the line was turned from 45° to 50° then

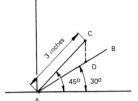

Figure 3.14. The isometric foreshortening scale

the amount of foreshortening would be more acute. Conversely if the angle was reduced the foreshortening would be less; however, the angle of 45° projected to 30° is the system generally used. For draughtsmen who regularly use foreshortened isometric projection it is possible to purchase from a good stationers or drawing office suppliers a ready-made isometric scaled rule. This of course eliminates the need for the construction of the scale on the drawing. When using this scaled rule, all dimensions on vertical lines are measured on the standard rule as before and all dimensions on lines angled at 30° to the horizontal, both left and right, are measured direct from the scaled rule.

So that the difference between these two methods of isometric projection, foreshortened and non-foreshortened, can be appreciated, the reader should now produce two drawings of the cube illustrated at Figure 3.12, one in foreshortened isometric and one in non-foreshortened isometric.

The ellipse in isometric

Let us suppose that we have a 3 in (76 mm) sided square in which there is a circle (Figure 3.15). Now if the square is drawn in isometric then the circle inside it will appear in isometric also. To produce the resultant ellipse we have to obtain a set of points on which the ellipse can be constructed. This can be done by various methods although two methods only will be dealt with here.

(1) Having first constructed the outline square in isometric (Figure 3.15a), the four mid-points where the circle touches the square A, B, C, D are drawn into the isometric square (Figure 3.15b). The four small squares where the diagonals touch the circle E, F, G, H can be transferred to the isometric view by the means already described (Figure 3.15c), and finally the ellipse outline can be drawn in (Figure 3.15d).

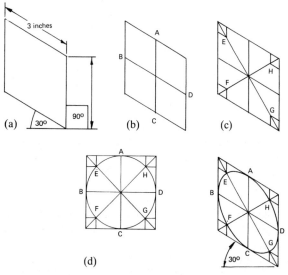

Figure 3.15. Constructing the isometric ellipse

(2) Another method of transferring a circle in orthographic projection to an ellipse in isometric projection is to draw a set of evenly spaced verticals or horizontals across the circle, and transfer the lines and the points at which they cut the circle to the isometric view (Figure 3.16). This method may be applied to any

OBJECTS IN THREE DIMENSIONS 41

Figure 3.16. Producing the irregular shape in isometric

irregular shape, and by measuring the distances along the lines (in this case the verticals) from a datum edge i.e. top or bottom, the resultant shape will be produced in isometric.

Exercises

Using both the methods described, produce ellipses in isometric from circles of the following dimensions.

3 in diameter	2½ in diameter
88 mm diameter	65 mm diameter
4 in diameter	3⅞ diameter

As will be apparent by now, the ratio of minor to major axis of the isometric ellipse is approximately 1:2. This means that the ellipse angle is 35° so that commercially only one template need be used. At this stage however, ellipse templates should NOT be used by the student; instead french curves should be used and persevered with to enable the student to produce the best possible standard of ellipse.

On the pages at the end of the last chapter are three exercises on isometric projection, the first being the simplest and the third the most difficult. The information for doing these exercises can be found in Chapter 2, in the section on first and third angle projection. The information in exercises No. 1 and No. 2 should be used to produce a plain or pure isometric view and the

42 OBJECTS IN THREE DIMENSIONS

information in exercise No. 3 should be used to produce a foreshortened isometric view.

Whilst still on the subject of objects in three dimensions we will take a quick look at two other methods of projection: oblique and axonometric. Both these methods are simpler than isometric but not as effective and are therefore not used as much.

Oblique projection

The oblique method is shown below. The front face of the object is drawn as it is normally viewed (the dimensions are then true

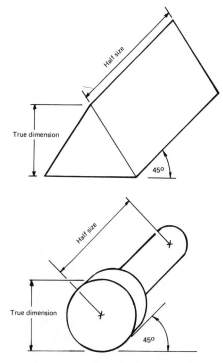

Figure 3.17. Oblique projection

and circles appear as circles) with all receding edges sloping at 45° to the right (Figure 3.17). With this method the receding edge is usually drawn to half full size (Figure 3.17).

Axonometric projection

With this form of projection the object can be rotated at any angle to the horizontal, but normally the angle of 45° is used (Figure 3.18). The plan of the object is drawn true with vertical

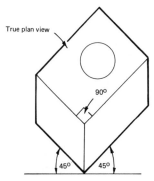

Figure 3.18. Axonometric projection

sides of the object drawn as verticals. Obviously circles in the plan would be shown as circles and not as ellipses as with other forms of three-dimensional pictorial representation.

Dimetric projection

Before finishing with this chapter on simple methods of illustration it is worth considering dimetric projection in similar brief manner to the way in which axonometric and oblique projection have been considered. Of all the 'draughtsmens' forms of pictorial representation, this probably produces the most sophisticated result, but is probably the most involved method. That is why draughtsmen in this country seem to prefer isometric of all the various methods. Dimetric however is used quite widely in the USA and, in brief, entails turning the axis of the illustration to an angle of 15° to the horizontal, the pictorial representation of the cube (Figure 3.19) being produced by projecting from plan and end elevations as shown. It will be noted that a certain amount of foreshortening occurs on all faces of the cube; none of the previous methods achieve this.

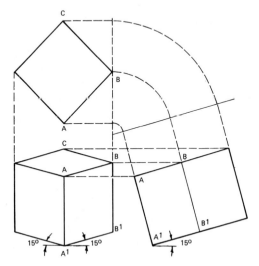

Figure 3.19. An example of dimetric projection

Note

There are many variations of angles and of foreshortening methods on all the many forms of projected three dimensional representations. I do not consider that the beginner need be concerned with them at this stage, and therefore the projections in oblique, axonometric and dimetric are only mentioned as comparisons of isometric and foreshortened isometric projections.

Exercises

Turn back to pages 20–22 of Chapter 2 and produce one pure isometric and one foreshortened isometric projection of each exercise.

4 A formula for technical illustration

Isometric, oblique and cabinet projections could all be described as 'draughtsman's pictorial representation'. They are methods by which illustrations can be produced in a mechanical or geometrical manner. In other words their true proportions are not considered one to the other. Their length is not decided on in relation to their height and width. They are laid out in a standard manner e.g. when an isometric drawing is laid out the object is laid at 30° to the horizontal, axonometric at 45° and so on and although these methods are used to portray mechanical subjects each method is only a compromise. A true technical illustration is not a compromise, but as good a likeness to the object as can possibly be given within reason. Such things as length, width, height, texture, light and shade, viewing position and perspective are taken into consideration when preparing the finished illustration.

Having worked through the previous chapters the reader will now realise that we have reached the point where elementary technical illustration can be considered. Bearing in mind therefore what has been stated in the opening paragraph of this chapter, how are we going to draw a better likeness of a given object than we would by using the methods previously mentioned? Primarily by studying the relationship of length, width and height of the object and drawing them in the correct proportions to each other. We must also consider any shape that is seen when either the length or the width or the height of the object is viewed. For instance let us consider the cube in Figure 4.1 with an ellipse drawn on each face. The height H is correctly proportioned to the length L and both H and L are correctly proportioned to the width W. Also the ellipse on each face is

46   A FORMULA FOR TECHNICAL ILLUSTRATION

correctly placed on its centre axis and the fatness or thinness of each ellipse is correct in relation to the others and in relation to the position from which the object is viewed. Compare now the cube in Figure 4.2 with the cube in Figure 4.1. It is tilted to a different angle and viewed from a different point, but its dimensions are unchanged, although the proportions of the ellipses have altered. Some ellipses have become fatter and some thinner although the whole object is still in correct proportion.

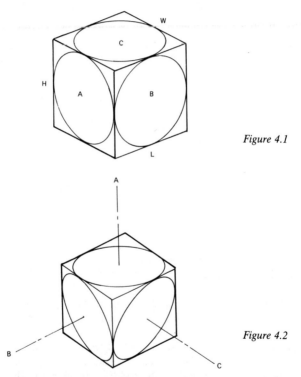

Figure 4.1

Figure 4.2

It is obvious, therefore, that the production of a correctly proportioned illustration is dependent on some basic means whereby the length, width and height can be proved to be in correct relationship with each other.

A FORMULA FOR TECHNICAL ILLUSTRATION   47

Consider now one end of a piece of round bar. If it is laid with its centre axis parallel to axis B (Figure 4.2) it would appear as indicated at B in Figure 4.3. Likewise if the bar were laid parallel to axis C (Figure 4.2) then it would appear as at C in Figure 4.3. Finally if its axis were turned vertical, parallel to axis 'A' (Figure 4.2), it would appear as at 'A' in Figure 4.3.

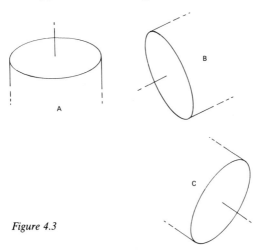

Figure 4.3

Now if we take diagrams A, B and C (Figure 4.3) and lay them one on top of the other (Figure 4.4) then we have some basis on which we can work toward true proportional illustration. This will become apparent as we progress.

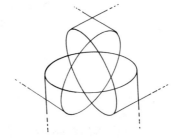

Figure 4.4

48   A FORMULA FOR TECHNICAL ILLUSTRATION

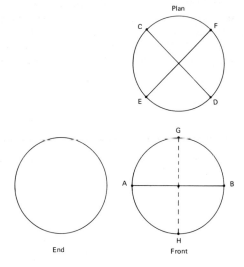

Figure 4.5

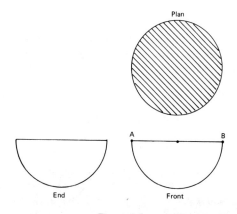

Figure 4.6

Slicing a sphere

Consider now a sphere, a football for instance. If a plan, front and end elevations of it were drawn we would finish up with three circles (Figure 4.5). Whichever way we look at a sphere we see an outline shape of a circle, therefore we can say that all points on its outline are the same distance from its centre point. Now if the sphere is sliced right across, through its centre point with a horizontal line A–B (Figure 4.5) then a plan, front and end elevation of the bottom half of the sphere will appear as at Figure 4.6. Now if the front elevation is tipped toward the viewer then the bottom half of the sphere will appear as at Figure 4.7.

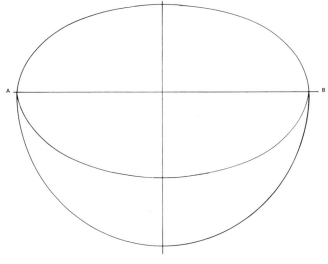

Figure 4.7

Likewise if we sliced our sphere vertically as at C–D (Figure 4.5), then the result would be as at Figure 4.8 and if a second vertical slice was made E–F (Figure 4.5) then the sphere would appear as at Figure 4.9.

What information then can be extracted from Figures 4.7, 4.8 and 4.9? We can say that in the plan view (Figure 4.5) the vertical slices through the sphere C–D and E–F are both equal in length. Compare now the relationship between the same lines

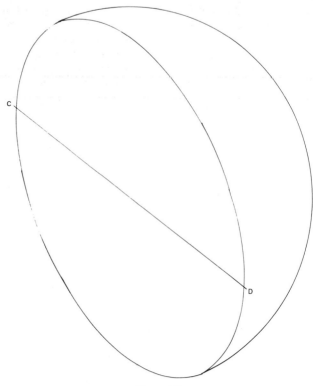

Figure 4.8

C–D (Figure 4.8) and E–F (Figure 4.9). These lines have now become foreshortened, both to different extents, but both are in correct proportion. As will be shown later the amount of foreshortening must vary depending on the viewer's position and the angle at which the object is turned toward the viewer.

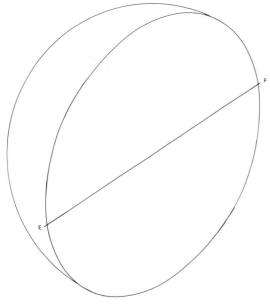

Figure 4.9

A foreshortening method

The student would probably do well to read over slowly and methodically the following paragraphs on foreshortening, after which some simple examples should be attempted.

In Figure 4.10 all the previous results obtained by slicing the sphere in one horizontal plane and two vertical planes have been laid one on top of the other. For instance the line A–B denotes the maximum extent of the horizontal slice through the sphere, and the lined ellipse the resultant shape. C–D denotes the maximum extent of the first vertical slice, and the broken lined ellipse the resultant shape. E–F denotes the maximum extent of the second vertical slice, and the chain dotted ellipse the resultant shape.

Any point on any of the three resultant ellipses is the same distance away from the centre point O but since the ellipses are all of different ratios, major axis to minor axis, then the distance away from point O will be foreshortened differently, but nevertheless, correctly. For instance if the diameter of the sphere were 8 in (203 mm) then line A–B would measure 8 in (203 mm) but line C–D is equal to a foreshortened 8 in (203 mm). Points C and

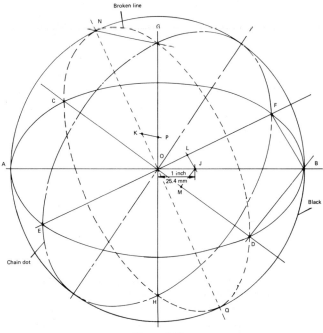

Figure 4.10

D still touch the outside of the horizontal slice through the sphere just as points A and B do. Similarly the line E–F is equal to a foreshortened 8 in (203 mm) but a different foreshortening than line C–D. Finally the line G–H is equal to a foreshortened 8 in (203 mm) dimension. It can be concluded from this information that we have now produced a correctly foreshortened length

A FORMULA FOR TECHNICAL ILLUSTRATION

of the sphere in line E–F, a width of the sphere in line C–D and a height in line G–H. Further, if A–B = 8 in (203 mm) then O–B = 4 in (101.6 mm) and O–F is a foreshortened 4 in (101.6 mm). Therefore if a line is drawn to link point B to point F, then any dimension along line O–B can be correctly foreshortened by projecting it parallel to line F–B onto line O–F e.g. line O–J = 1 in (25.4 mm) and line O–L = 1 in (25.4 mm) foreshortened LENGTHWISE. The same point J, on line O–B projected on to line O–D parallel to a line drawn from point B to point D, would give a line O–M which would represent 1 in foreshortened WIDTHWISE.

Referring now to Figure 4.5 in the front elevation, the height of the sphere is indicated by G–H, therefore they can be shown as in Figure 4.10 in a foreshortened form. We link the true dimension of 8 in (203 mm), indicated by line N–Q, to the foreshortened dimension, indicated by line G–H, by drawing in line N–G. Now if 1 in (25.4 mm) is measured along line N–O linking O–K and is projected to line G–O at point P, parallel to N–G, then O–P is a foreshortened 1 in (25.4 mm) in height. We can now use this basis to produce correct foreshortening of length, width and height.

Using the foreshortening method for length and width

With centre point O, draw a circle 8 in (203 mm) in diameter to represent an 8 in (203 mm) sphere. Now construct an ellipse by the trammel method, making the major axis A–B 8 in (203 mm) and the minor axis C–D 3 in (76.2 mm) as in Figure 4.11, bearing in mind that the axes must cross at 90° at point O. This ellipse represents the resultant shape after a horizontal slice through the sphere has been made.

Now draw a line through the ellipse at any angle, preferably at this stage between 25° and 35° to the horizontal, to cut through centre point O. This line must cut the ellipse at point E, pass through point O, and cut the ellipse again at point F. This will form the angle at which an illustration is laid out.

Next draw two short lines parallel to E–F that will just touch the sides of the ellipse (tangents), call the points where these lines touch the ellipse G and H, then link G and H with a line through the centre point O. We can now use lines E–F and G–H for our foreshortened length and width respectively. We have now produced a proportional grid on which to work.

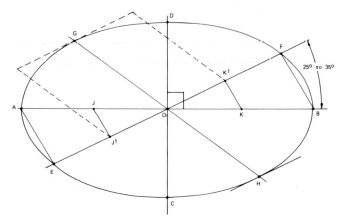

Figure 4.11

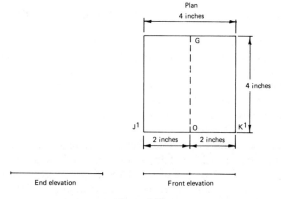

Figure 4.12

Let us now consider a square of 4 in (101.5 mm) side (Figure 4.12). The length and width can be seen in the plan and as shown in the end and front elevations it has no height. To produce a drawing of this square on our proportional grid, first measure 2 in (50.8 mm) from point O on line O–A and 2 in (50.8 mm) from point O on line O–B, to give points J and K (Figure 4.11). Points O and G are the mid points of the front and back of the square respectively (Figure 4.12). Link A and E and then B and F (Figure 4.11) and draw lines parallel to them from points J–J^1 and K to K^1. J^1 to K^1 is the foreshortened length of the front face of the square J^1, O, K^1 (Figure 4.12), point O being the mid point of the line. In Figure 4.11 line A–B is 8 in (203 mm) long, and A–O is therefore 4 in (101.5 mm) long; as line O–G is a foreshortened version of O–A, then by drawing a line through point G parallel to line E–F, and linking it with lines from points J^1 and K^1, parallel to line G–H, we have a proportional drawing of the square from Figure 4.12.

Examples

Now draw the plan, front and end elevations of a square of 5 in (127 mm) side that has no height, then produce a foreshortened drawing of the square placing the mid point of the front face, point O, on the centre of the grid and following through the method described in the preceding paragraphs. When this is complete do the same with a 3 in (76.2 mm) sided square also of no height.

Foreshortened height

Having dealt with a square which has length and width we can move on to consider the third dimension, that of height. Referring now to Figure 4.10, in conjunction with the last paragraph of the section headed 'A foreshortening method', it is apparent that our foreshortened height will be obtained by the same means.

56 A FORMULA FOR TECHNICAL ILLUSTRATION

Refer now to Figure 4.13 in which the base of a 4 in sided cube has been completed. To determine the height it is necessary to draw in the ellipse having a major axis of L–M at 90° to the axis E–F again about the centre point O and passing through the pre-determined 'width' points of G and H.

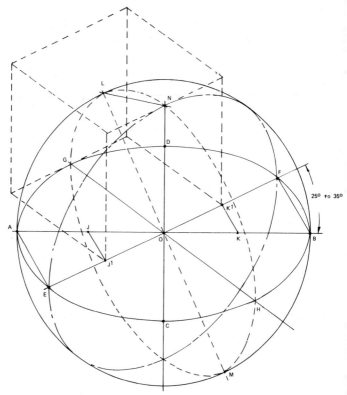

Figure 4.13. Constructing the cube on the proportional grid

To do this a trammel should be made and laid along the major axis L–M and the length L–O marked off on it (Figure 4.14). Now the major axis is known and marked but the minor axis is unknown. We know that the ellipse must pass through points G

A FORMULA FOR TECHNICAL ILLUSTRATION

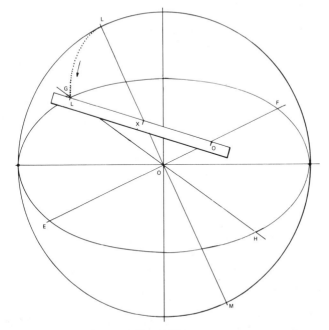

Figure 4.14

and H, therefore if point L on the trammel is placed on point G on the horizontal ellipse and point O on the trammel is placed on the axis line E–F, then point X can be marked on the trammel where the trammel edge crosses the major axis L–M. Keeping the trammel moving in the direction indicated by the arrow in Figure 4.14, point X sliding along L–M, and point O on the trammel sliding along E–F, a vertical ellipse can be produced. Now referring to Figure 4.13, where this ellipse cuts the centre vertical we will call point N. If L is linked to N then O–N is a foreshortened HEIGHT of our cube. Point N is the mid point of the top front edge, directly above point O, the bottom front edge, and therefore the outline of the cube can be linked through these points as indicated in Figure 4.13.

58 A FORMULA FOR TECHNICAL ILLUSTRATION

Examples

Referring to the preceding paragraphs construct two cubes, one of 6 in (152 mm) and one of 203 mm side.

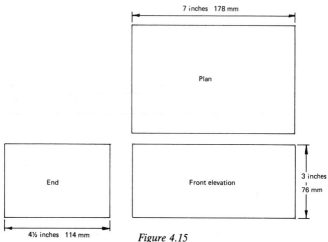

Figure 4.15

Let us now consider the construction involved in the drawing of a brick of length 7 in (178 mm), width 4½ in (114 mm) and height 3 in (76 mm), as shown in the orthographic drawing Figure 4.15. The brick is 7 in (178 mm) long. First draw a 7 in (178 mm) diameter circle to represent the sphere and draw in it, with the aid of a trammel, a horizontal ellipse to represent a horizontal slice through the sphere. Remember that the fatter this ellipse is drawn the more will be seen of the top of the brick when completed. The thinner this ellipse is drawn, then the less will be seen of the top of the brick. With this in mind let us for this exercise make the minor axis of the horizontal ellipse 4 in (102 mm) (Figure 4.16). As with the previous exercises we now have to decide upon an angle at which to lay the drawing. Previously we laid it at any angle between 25° to 35° to the horizontal (Figures 4.11 and 4.13). For this exercise we will lay the brick at 20° to the horizontal (Figure 4.17) and we will use similar letter annotations to those in the previous chapter.

A FORMULA FOR TECHNICAL ILLUSTRATION

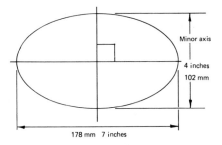

Figure 4.16

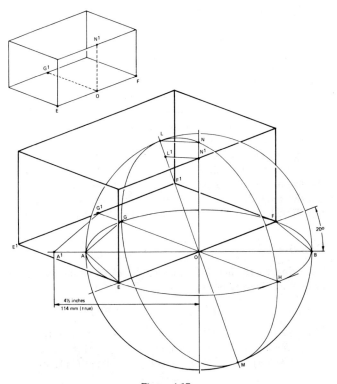

Figure 4.17

Line A–B is 7 in (178 mm) long, so the point where the line laid at 20° to the horizontal cuts the horizontal ellipse at points E and F represents the extent of the foreshortening of line A–B. As with previous examples we will lay the front base line along line E–F and again we will call point O the centre point of the front base line. To obtain the width of 4½ in (114 mm) the 'true length line' A–B must be extended through point A and the distance of 4½ in (114 mm) measured from point O (centre front base) through a point A to give point A^1. Line G–H must now be extended through point G and point A^1 projected onto this extension at point G^1, parallel to a line drawn from point A to point G. Therefore distance O–G^1 is the foreshortened width of 4½ in (114 mm). Lines can now be drawn parallel to O–G^1 from points E and F to cut a line drawn through point G^1 drawn parallel to E–F at E^1 and F^1. The resultant shape will be the base of our brick (Figure 4.15). As before, to obtain the correctly foreshortened height, a second ellipse must be constructed; and as before, O–L is a true dimension and O–N is the foreshortened version of this dimension. Now O–L is 3½ in (89 mm) in length but the height we require is 3 in (76 mm), therefore 3 in (76 mm) must be measured from point O and marked off at L^1. This point can then be projected back onto line O–N (parallel to a line drawn from point L to N) at point N^1. It will be seen from the thumbnail sketch (Figure 4.17a), how, by drawing parallel lines through the points G^1 (centre, back, base) and N^1 (centre, top, front), the outline shape of the brick can be linked up.

Exercise

Referring back when necessary to the explanation just given, draw the following bricks:

Height	*Length*	*Width*
2 in (51 mm)	5 in (127 mm)	4 in (102 mm)
5 in (127 mm)	6 in (152 mm)	3½ in (89 mm)
3½ in (89 mm)	7½ in (191 mm)	4½ in (114 mm)

Lay them at any angle to the horizontal between 15° and 40°.

It will become apparent as the student becomes more familiar with proportional illustration that not all illustrations require the construction of the three ellipses. The bricks just dealt with only required the constructing of the horizontal ellipse for length and width and one vertical ellipse for height. In some of the later exercises only two ellipses are required, in others all three ellipses are necessary to produce a correct illustration.

If there is any difficulty in finding where the points are taken from for the three ellipses, then turn back to the section headed 'A foreshortening method' and go over it following through the construction of the solid, broken and chain dotted ellipses of Figure 4.10. When completing the exercises the student should use french curves as guides for drawing in curves, ellipses etc. Ellipse guide templates should not be used at this stage. Attention should be given now to quality and accuracy of the finished illustration, weight of line and general pencil finish. The student may well find that to produce an illustration to the required standard in pencil, he/she may prefer to draw it out with a 3H or 4H grade of pencil, and then to complete the illustration with a softer grade. It is worth experimenting to find out which grade suits the individual best. Outlines should be heavier than corner lines and edges (changes of plane lines). As a guide to this the student should refer back to the first chapter 'What is technical illustration?', examining the line drawings and observing where light lines and where heavier lines are used.

Proportional illustration

When using the grid remember that the viewpoint in the illustration is obtained by a combination of a certain two points:

Point A is the proportion of the minor axis to the major axis in the horizontal ellipse.

Point B is the angle at which the object to be illustrated is laid to the horizontal.

If the angle is steep $Y°$ and the horizontal ellipse is a fat one, i.e. the minor axis is longer in proportion to the major axis, then

62   A FORMULA FOR TECHNICAL ILLUSTRATION

a great deal of foreshortening is going to take place (Figure 4.18). If however, the horizontal ellipse is shallow and the angle X°, at which the object to be illustrated is laid, is narrow then the opposite will happen (Figure 4.19). This grid can be drawn up to

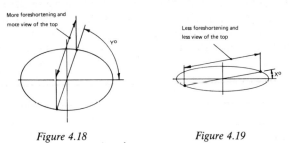

Figure 4.18   Figure 4.19

suit whatever viewpoint may be required. Sometimes the illustrator will want to see down into an object; sometimes more of the front and less of the end of the object will be required. It is necessary, therefore, to construct the grid to give the best possible viewpoint of the object to be illustrated. The ability to do this will come with practice and experience.

Producing a simple illustration on the grid

Before commencing with the next set of exercises let us take the simple bracket shown at Figure 4.20 and, scaling from the orthographic drawing, we will work our way through, stage by stage, to the completion of a pencilled line illustration. Inking and shading will be dealt with later.

After making a preliminary three-dimensional sketch to clarify the shape and proportion of the bracket we can draw up the grid. As it is a simple object and has no features of any real complexity a grid of only two ellipses giving a general viewpoint will suffice (Figure 4.21). We will look at the bracket from the direction of the arrow in Figure 4.20 (PLAN VIEW) and we will lay the point A on the centre of the grid for ease of positioning.

A FORMULA FOR TECHNICAL ILLUSTRATION 63

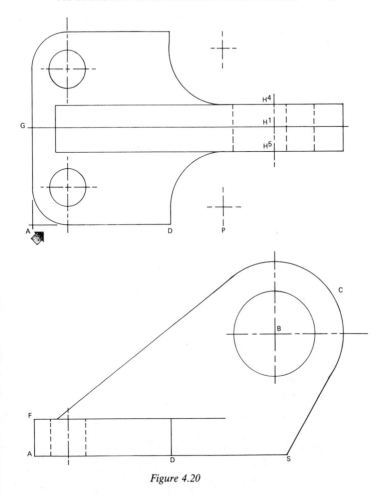

Figure 4.20

Figure 4.21 shows the grid drawn up in a manner already described with the vertical ellipse passing through points (1) and (2). Looking in the direction of the arrow in Figure 4.20 (PLAN VIEW) we will see the hole through the web of the bracket (B) and the radius (C) front elevation. To work out their correct proportions this particular vertical ellipse is required and not the other one.

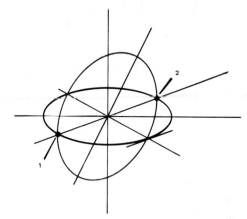

Figure 4.21

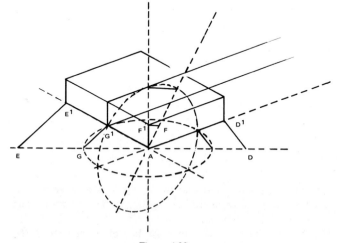

Figure 4.22

A FORMULA FOR TECHNICAL ILLUSTRATION 65

Figure 4.22 shows how points D (to the right of A), E (to the left of A) and F (directly above A) are marked off, using the dividers, from the orthographic drawing (Figure 4.20), placed on the *true* dimension lines and finally projected to the foreshortened lines to give correct length $A-D^1$, width $A-E^1$ and height $A-F^1$, as previously described. Extensions of points $G-G^1$ give the centre lines of the bracket through the base and along the plate top.

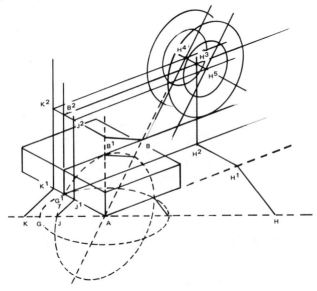

Figure 4.23

The distance of point H from point G (Figure 4.20) is next measured with the dividers and marked out from A to H (Figure 4.23). Likewise the height of the hole centre B (Figure 4.20) is measured with the dividers and marked off on the *true* height line A–B (Figure 4.23). By projecting, via the foreshortened height, line B^1 is obtained and then projecting the extended front centre line from G^1 we obtain B^2. The centre line from point B^2 is extended back and the points H^1 and H^2 are projected from H. A vertical from H^2, intersecting with the extension of B^2, gives us H^3, the point at the centre of the hole. In similar manner points H^4 and H^5 are made by extending through H^3 to the foreshortened width of the web from A–J, A–K *true* dimension and projecting via J^1, K^1 thence to J^2, K^2. Finally ellipses can be drawn as shown in Figure 4.23.

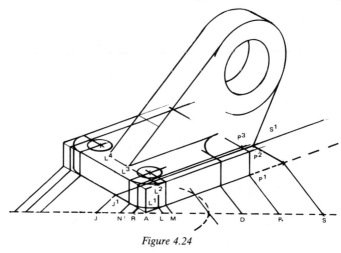

Figure 4.24

Referring to Figure 4.24, A–L on the *true* dimension line gives us the distance from the plate face at which the web joins the plate top end, and is arrived at via points L^1, L^2 to give points L^3 and L^4. Lines are then drawn to 'blend' with the outer radius ellipses. The centre of the two holes in the plate are measured from the edges A–M and A–N and projections made to give their centres. On completing these ellipses the blend radius of D–P is produced to P^3 at the intersection with the projection of point R from A–R (*true* dimension line). A–S (Figure 4.20 (FRONT VIEW)) will give the point at which the end of the web joins the horizontal; in Figure 4.24 J^1 is extended and the intersection is at S^1.

If any difficulty has been experienced whilst working through this bracket, don't go on to the exercises but go back and work through the previous text on 'A foreshortening method'. Remember that the ONLY dimensions on the illustration that are true dimensions are the major axes of the ellipses; lengths, widths and heights must ALL be foreshortened.

Figure 4.25

The fifth stage in the production of our illustration is that of 'lining in' and it is the most rewarding stage when the illustration is made to come to life after all the donkey work applied in the earlier stages. As shown in Figure 4.25 outlines are heavied-up; change-of-plane lines, corners and edges remain light. Holes have light shadows applied to help the three-dimensional effect. Two different types of shadow are shown here.

Intersections

An intersection of two cylinders is shown in Figures 4.26 and 4.27. In Figure 4.26 draw two centre lines perpendicular to each other for the axis of the two cylinders. Draw the ellipses equal to the size of the cylinders. Select points along the vertical centre line of one of the cylinders and lay off equal distances. Measure the same distances along the vertical centre line of the other cylinder, then draw lines from the points on the vertical centre lines of the cylinders to the outer edge. From these points draw lines parallel to the cylinders until they intersect at A, B, C, D, E and F. Connect these points with an irregular curve to make the proper line of intersection.

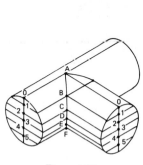

Figure 4.26

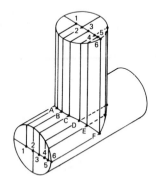

Figure 4.27

Follow the same procedure for Figure 4.27 but divide the horizontal centre lines of the cylinders into equal points.

The method for drawing the intersection of two cylinders at an angle less than 90° is shown in Figure 4.28. The top and front orthographic views of the intersection are shown and the methods for making the illustration. Selected points are laid off on the orthographic views as indicated by H, N, R, X on the top view and A, B, C, D, E, F, G on the front view. These points are located along the proper axis of the illustration and are connected by using an irregular curve to form the intersection.

A FORMULA FOR TECHNICAL ILLUSTRATION

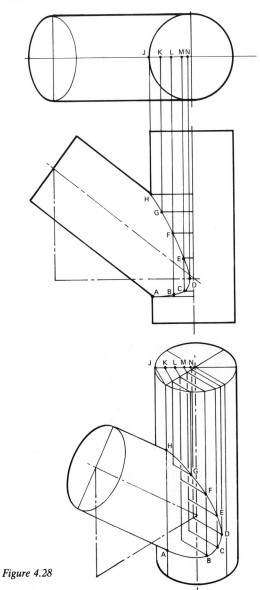

Figure 4.28

Exercises

4.1 Redraw the two given orthographic views and complete the third view laying them out in first angle projection. From this informaton a three-dimensional proportional drawing should be constructed.

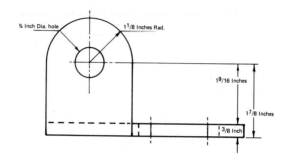

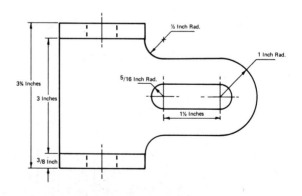

Exercise 4.1

A FORMULA FOR TECHNICAL ILLUSTRATION 71

4.2 Redraw the two given orthographic views and complete the third view laying them out in third angle projection. From this information a three-dimensional proportional drawing should be constructed.

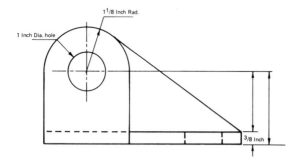

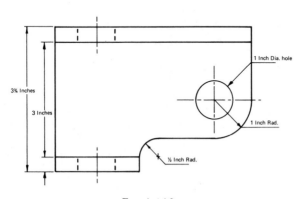

Exercise 4.2

72 A FORMULA FOR TECHNICAL ILLUSTRATION

4.3 From this sketch produce an orthographic drawing laid out in third angle projection. From this information produce one three-dimensional drawing in foreshortened isometric and one proportional drawing in direction of arrow X.

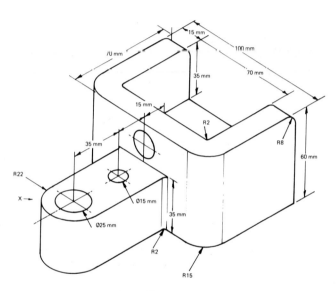

Exercise 4.3

A FORMULA FOR TECHNICAL ILLUSTRATION 73

4.4 From the three views given produce a proportional drawing looking in direction of arrow X giving less view of the top and more view of the front and side.

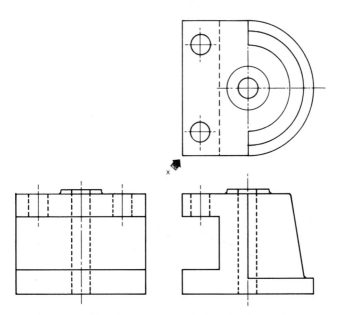

Exercise 4.4. Third angle projection. Scale from this drawing twice up to produce a proportional drawing

4.5 From the sketch given produce an orthographic drawing laid out in first angle projection from which a proportional drawing looking in the same direction as the sketch given should be constructed.

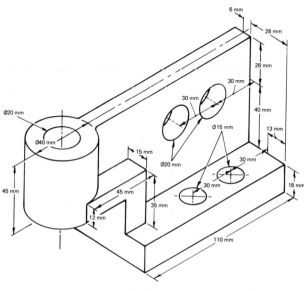

Exercise 4.5

5 An introduction to perspective

For those who wish to delve deeply into the whys and wherefores of perspective theory there are a number of good books available. Whether or not it is necessary for a student learning to produce simple technical illustrations to be able to plot and project the shape of objects inclined to the picture plane, for instance, and then to be able to produce the shadows cast by a light source from a given point is, to my mind, debatable! As a student, I, like others, had to 'go through the mill' to pass certain examinations but having passed in the subject I never once used 'perspective theory' again when producing illustrations commercially. If, after all, a person calling him or herself a technical artist cannot put the right amount of perspective on an illustration by eye, having constructed the illustration by the method already described, then a pre-draw matrix called a perspective grid would be used. After laying tracing paper over the grid the measurements and perspective lines are used as a guide in constructing the three-dimensional illustration. In our world 'time is money' and no client pays the illustrator for 'non-productive time'. If then the illustrator can produce an illustration to a standard acceptable to the client by some short-cut, then economics alone will dictate whether a theoretical method that takes much construction time is used or not used. Having made these points, I would hasten to point out that even to produce simple technical illustrations to a satisfactory standard one must have a certain understanding of some elementary perspective principles.

We are all aware that if we stood on a flat plain and gazed at the horizon then all that we could see would recede into the distance; and the further away one gets from fixed objects, the smaller and closer together they appear to be. If there were a road on this flat plain, the further it receded the narrower it would appear to be, until eventually it merged into one point (the 'vanishing point') on the horizon (Figure 5.1). The presence of buildings on the roadside demonstrates this point more clearly

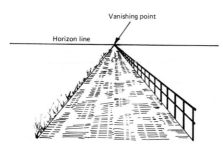

Figure 5.1. Centre viewpoint

(Figure 5.2). The roofs of the buildings are higher than we are and are all at different heights from the ground. Consequently they all have different vanishing lines, but since the front of each building in Figure 5.2 is parallel to the kerb, they all vanish to the same vanishing point. If we moved to left or to right or jumped up and down then the view would obviously be different. This then is what is meant by single point perspective: all vanishing lines converge on a single point on the horizon.

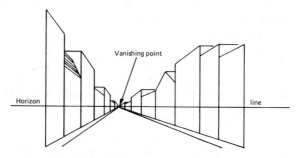

Figure 5.2. Left of centre viewpoint

AN INTRODUCTION TO PERSPECTIVE 77

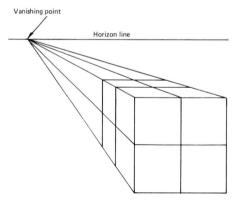

Figure 5.3. The cube in single point perspective

Figure 5.3 shows a cube drawn in single point perspective. This time we are above the object and therefore looking down onto it. Its length diminishes with its distance from the viewer and subdivisions on its surface will consequently vary in area on its top and side surfaces compared to the equal surface areas on its front face. The horizontal and vertical dimensions are correct, of course, whilst those on the top and side get less as they go further back. Single point perspective is rarely used in technical illustration as it does not give pleasing proportions to mechanical shapes.

If we looked down on an object and observed that its vanishing points were as indicated at Figure 5.4, then we would

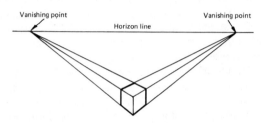

Figure 5.4. The cube in two point perspective

78 AN INTRODUCTION TO PERSPECTIVE

call this two point perspective, as its vanishing lines are directed to vanishing points to left and to right of our 'station point'. All of these situations are used in industrial art; architectural illustrators being the users of the widest of perspective variation. For instance the architectural illustrator may need to use what the Americans call a 'bugs eye view', a viewpoint taken from ground level and observing the scene as indicated as at Figure 5.5. A new town centre development may on the other hand

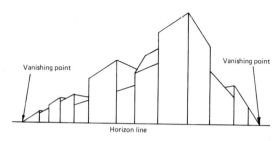

Figure 5.5. Ground level viewpoint

require an illustration from an aerial viewpoint: see how the perspective is almost exaggerated and uses three vanishing points in Figure 5.6.

Technical illustrations are usually produced in three point perspective. Edges of the object vanish to left and right also below the object but we normally take a viewing position slightly above the object. The three-dimensional object we used to describe orthographic information can be looked at again to view three point perspective. Fold the object up as you did at Figure 4.2 but this time place it on a flat surface so that you are directly in line with the corner as indicated at Figure 5.7. If we were to draw this object in three point perspective the vanishing points would be as indicated here. If we move our station point (the position from which we view the object) above, below, to left or to right of our existing position we will of course see a different picture, but the closest illusion on paper that we, as artists, can produce is with three point perspective. Whichever point we choose to view the object from, three point perspective is

AN INTRODUCTION TO PERSPECTIVE

guaranteed to give the best likeness acceptable to the humen eye, but it is of course the most difficult to complete effectively.

The majority of readers will be aware that if we wish to find the centre of a square or a rectangle without resorting to measuring, then by drawing diagonals the problem is easily

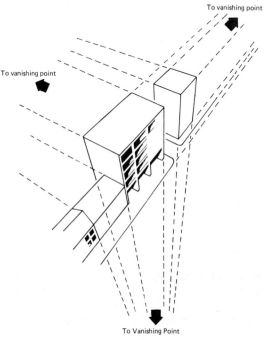

Figure 5.6. Perspective using three vanishing points

resolved. A simple exercise is shown at Figure 5.8 which uses the diagonal method together with knowledge already acquired to further clarify foreshortening in perspective. The subject is a simple rectangular plate, but since our viewpoint gives a resultant illustration that has exaggerated perspective, then as you work through the exercise the foreshortening and the vanishing lines will produce an acute perspective result.

80   AN INTRODUCTION TO PERSPECTIVE

In Figure 5.8, 2–7 is ½ of 2–3 and 6–7 is ½ of 3–4.
a. Find the midpoint line 7–6 and draw in vanishing lines.
b. Find the midpoint of line 2–7.
c. Find ⅛ point of line 7–6.
d. Find midpoint of line 7–3.

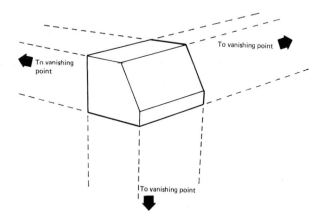

Figure 5.7. The cut-out object used to observe its perspective vanishing lines

As has already been pointed out, the aim of this book is to cover the study of simple technical illustration, and the level of work therefore requires the student to acquire the ability and know-how to the standard where simple mechanisms, valves, pumps and other engineering subjects can be drawn correctly and convincingly. Since these subjects are generally small in size very little perspective can be 'built-in' to the resultant illustration. If a small object is given too much perspective then it is just as bad as none at all. The amount of perspective applied at this level of work is therefore so small that it is best applied by eye but based on the points already made. With reference to Figure 5.9, a cube, of sides measuring two units – in this case two inches – has been constructed following the previously-described formula, i.e.:

AN INTRODUCTION TO PERSPECTIVE 81

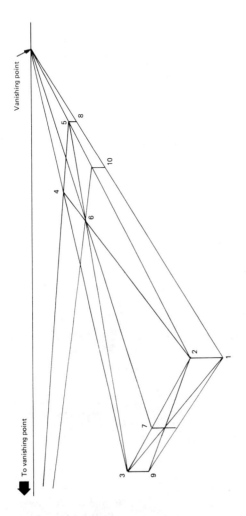

Figure 5.8. See text questions on diagonal measurement

AN INTRODUCTION TO PERSPECTIVE

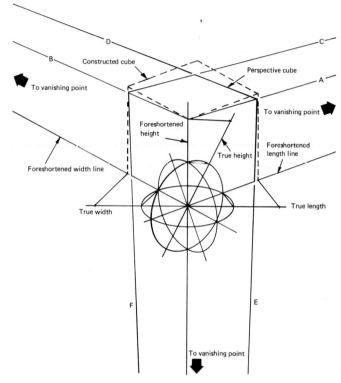

Figure 5.9

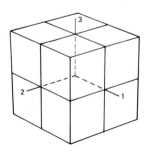

Figure 5.10. The 'perspectively corrected' cube

(1) Correct length, width and height are marked off and projected onto their appropriate foreshortening lines.
(2) The cube, indicated by the broken line, is completed with parallel lines.
(3) The base of the cube lying on the foreshortened length line and the foreshortened width line are left untouched as is the edge nearest to us, the foreshortened height line.
(4) The amount of perspective is assessed by eye and the vanishing lines A and B are drawn in.
(5) Where line A cuts the broken line of the cube the rear vanishing line to left is drawn in, line D.
(6) Where line B cuts the broken line of the cube the rear vanishing line to right is drawn in, line C.
(7) Lines E and F are drawn in to complete the outline of the perspective cube.

Note how the top corner of the grid has moved down: this is of course the culmination of the top sloping down to left and right as far as our eye is concerned, and will vary with the amount of perspective that is put on, as well as with the angles that the grid is drawn at. This grid is the same as that used at Figure 4.21, but the third ellipse has been drawn so as to 'prove' its correctness to the reader.

Figure 5.10 shows the 'perspectively corrected' cube with subdivided surfaces, the subdivision lines now being in three point perspective with the cube. Any axis lines drawn through the centres of these surfaces would also be in three point perspective and not parallel to the originally drawn grid lines e.g. the lines distinguishing the base to left and right of the cube's base and the vertical line of the corner nearest to the viewer. Draw up your grid at any angle and following through the description draw a 'perspectively corrected' cube yourself. When you start illustrating, apply the lessons learned on perspective together with your other experiences learned in the other chapters. Don't just think of perspective as a 'non-applying' subject; make it apply to your illustrations depending on their size and shape.

6 Basic techniques

The detail contained in a technical illustration is most important. If the illustration is to communicate well it must look 'life-like'. If, as has already been discussed, the artist cannot use colour or tone, and as shading must be kept to the absolute minimum, then the small amount of shading that is used must really count and details like radii, screw heads, nuts and welds must all look convincing.

The purpose of this chapter is to give some examples of the main features that go into most illustrations. It is of course impossible to cover all these features but the following drawings and comments should help to clarify a lot of the objects that make up the illustrations. These are 'tips' if you like but of course they are not intended to be the ONLY way of illustrating set items; for instance you may shade a thread in a certain manner on one illustration, but when the same method is used in a different situation it may be found that it doesn't work so well and has to be modified.

Radii

Let us start with something simple, the blend radius for instance. Whether this is external or internal we always treat it with irregularly broken lines describing where the radius begins and where it ends (Figure 6.1). We draw the corner of the object first, then put in the radius (Figure 6.2) and finally remove the corner lines (Figure 6.3). In certain instances it works better if a few lines are drawn around the radii but this is not always

BASIC TECHNIQUES

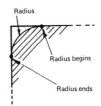

Figure 6.1. External radius in plan view (section)

necessary. The ceiling rose (Figure 6.4) shows clearly the radius's beginning and end lines. Note the break in the radius to indicate the highlight. The light switch (Figure 6.5) has a very small radius around its front edge. The bottom left has been illustrated with three lines to give a little weight to an otherwise white drawing, while the radius lines are faded out to the right.

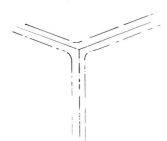

Figure 6.2. Corner first, then radius, finally remove corner

Figure 6.3. Corners removed and radius applied

Figure 6.4. Plastic ceiling rose

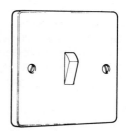

Figure 6.5. Plastic light switch

On the extreme right we are looking across the radius and so see no lines at all except the line formed by the 'edge' of the radius (Figure 6.6).

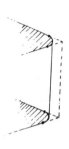

Figure 6.6

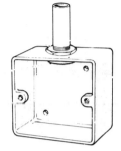

Figure 6.7. Termination box

The electrical termination box at Figure 6.7 is a good example of an internal radius using two irregularly broken lines. This illustration has a little light tone to help on its shadow side but could equally well have had line shading on it.

Threads

Threads of one form or another are probably the most common part of technical illustrations. An internal and external vee form thread is shown at Figure 6.8. Threads are one subject where we have to exaggerate to give a convincing result. They always look best if they appear shiny and newly cut with the main shadow of the external thread on the inside at the top, while the shadow of the external thread is heaviest at the bottom; the position of the highlights are also opposite to each other.

Threads of very small outside diameter present certain problems, as they are too small to shade in the normal manner to make convincing. They are best treated as in Figure 6.9, leaving the left side (the light side) as a highlight and lining in the right side (shadow side). Then rule a feint pencil line down just inside the right edge and dot the shadow cast by each thread onto the one below.

Figure 6.8. An internal and an external thread

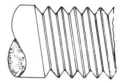

Figure 6.9. Illustrating a small screw thread

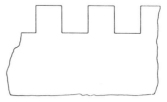

Figure 6.10. The shading of both external threads and surface of a split type collet

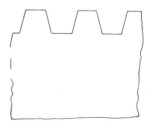

Figure 6.11. Square and Acme threads are simplest to indicate by a series of discs to represent the outside and root diameters

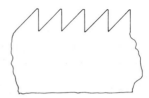

Figure 6.12. Basic thread forms – from the top they are: the basic vee thread, the square thread, the Acme thread and the buttress thread

Large areas of thread, e.g. large outside diameters, mean that we can give more separate areas of shade and highlight as shown in Figure 6.10. The collet has three splits down its axis and is used on a variety of lathes in the engineering industry. In addition to the thread shading, note also the line shading slightly lighter on the top than on the bottom helping to give the impression of a fine and smooth surface finish.

The main non-vee form threads that we come across, often when illustrating machine tools for instance, are the Buttress, Square and Acme forms as a comparison with the standard vee form. The method usually adopted by the student to indicate these threads is to draw a set of 'discs' and to treat them as shown in Figure 6.11.

Screws, nuts and bolts

Figure 6.13 shows five of the most commonly used types of screw, The bolt has a length of plain shank and hexagon head the same as the set screw and the same as the standard nut. All three, setscrews, bolt and nut are champhered on the top, and are best illustrated with one flat directly facing the viewer, as indicated by the nut. The socket head is similar to the cheese head but is normally shown with a parallel knurl on its head and with a small champher on the top. The method of tightening the socket head is with an 'Allen' key, a key that is hexagonal in section. The female end must therefore be shown in the top of the socket head screw.

Three simple stages in drawing the standard bolt are shown at Figure 6.14. The outside diameter of the shank is, for illustration purposes, about the same dimension as the distance across the flat directly facing, being about one half of the outside diameter of the head. When drawing the nut, however, it is necessary to 'open-up' a little the inside diameter to create the illusion that it fits the thread (Figure 6.8 – internal and external conduit threads).

The castle nut shown here is used in conjunction with the split pin, which is inserted through a hole drilled through the shank

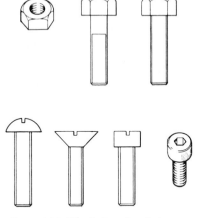

Figure 6.13. Simple fastening devices encountered by the technical illustrator, these are the nut, bolt and setscrew and round head, countersink, cheese head and socket head screws

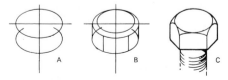

Figure 6.14. Three stages in drawing the standard bolt

Figure 6.15. The castle nut and split pin, the round head and countersink head screws

90   BASIC TECHNIQUES

onto which the castle nut has been tightened. The pin then stops the nut from vibrating loose after the ends have been opened.

The other three screw heads are easier to draw. The round head has a domed top to it so we must use half a fatter ellipse for its top. The socket head normally has a parallel knurl, but don't draw it all the way round. If the drawing size allows it, put a little knurling on the left but more on the right (the shadow side). If the size doesn't allow this treatment, keep it for the right side only, and make sure in any case that the knurl lines grow wider apart as they come around the circumference (Figure 6.15).

Washers

The three main types of washers are shown at Figure 6.16 and are as follows: (a) plain washer (b) spring washer (c) tab washer.

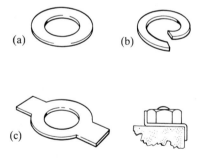

Figure 6.16. Three different types of washer – the plain, the spring and the tab. The tab washer is shown before and in use

The spring washer is flattened under the nut as it is tightened and so prevents the nut from vibrating free, as does the tab washer using a different principle, as is shown in Figure 6.16.

BASIC TECHNIQUES

Bearings

The five main types of bearings are shown at Figure 6.17.
(a) The ball bearing carries combined radial and axial loads.
(b) The cylindrical roller bearing carries radial loads only.
(c) The needle roller bearing will fit into smaller spaces than (b) but performs the same function.
(d) The taper roller carries high radial loads and axial thrust in one direction only. When showing bearings in an illustration we never section them unless they are hiding some important feature of the illustration. You should show their external features only. Don't show them with their 'skins' peeled back as here these views are for clarification only.

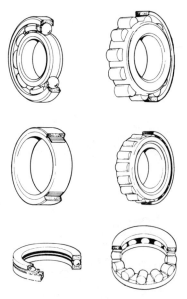

Figure 6.17. A variety of bearings. The left column shows the ball, needle and ball thrust bearings while on the right are the roller, taper roller and roller thrust bearings

Circular objects

The lathe chuck shown in Figure 6.18 has been given minimum amount of shading, just enough to help show its circular external shape and the internal diameter through its central axis. The shading is around its circumference and about its centre axis.

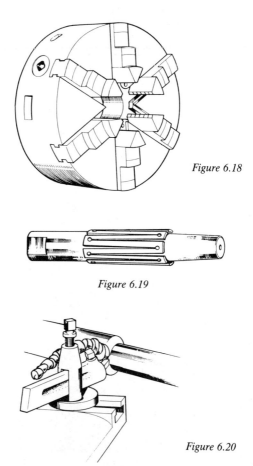

Figure 6.18

Figure 6.19

Figure 6.20

Lines parallel to the centre axis can also be used. Figure 6.19 shows a fine, smooth surfaced expanding mandrel with clusters of line shading that have been broken to describe the circular surface. Note also the vertical line shading on the driving flat.

Finally, take a look at the different circular objects that make up the illustration of the lathe toolpost in Figure 6.20. The metal being turned is shaded with clusters of line shading, centre axis parallel. The swarf that curls up over the tool bit is line shaded around its circumference, while the tool post and dished washer are shaded vertically.

Line shading on flat surfaces

The group of small mechanical objects in Figure 6.21 give a good indication of a variety of line shading. They are all of a machined finish and so a small amount of shading only can be used to help make them look convincingly three-dimensional.

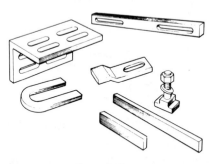

Figure 6.21

When applying any shading it is best to use the shading to register a contrast between one surface and another surface, as here no two adjoining surfaces have been treated. Note also the discreet use of perspective even on such small objects, together with slight shadows to give the desired effect.

94   BASIC TECHNIQUES

When two or more smooth flat surfaces of different finish are encountered, the illustrator has to be even more careful to give the right impression. The demagnetiser shown in Figure 6.22 has

Figure 6.22. The shading on a highly polished surface, here on the top plates of a demagnetiser

a smooth surface all over but the top two plates are polished, hence the vertical reflection lines as opposed to the perspective lines on the end face.

Stipple shading

Stipple is applied as a series of dots with the tip of the pen and is used to indicate a number of different surfaces. Figure 6.23 shows how stipple in varying weights is used to show a broken surface.

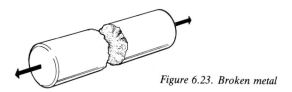

Figure 6.23. Broken metal

The cast iron bracket in Figure 6.24 is stippled over most of its surface but has strips of highlight left white on circular surfaces. It is shaded up to its edges in an irregular manner in order to make a contrast against the white background.

The extended sections of Figure 6.25 indicate a further use for stipple. Here the shape of the section and its clarity to the viewer are of prime importance, and since three dimensions – length, width and height – are better than two dimensions to describe a

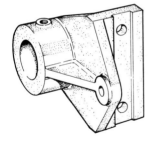

Figure 6.24. A rough cast surface indicated with stipple

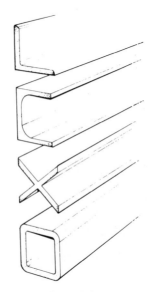

Figure 6.25. The shape of extruded sections is emphasised by stipple

shape, only the ends of the extrusions are shaded. Note the use of two point perspective, the use of shadows and indication of edges and radii. Observe how the stipple is slightly heavier at the edges of the section to give a strong contrast against the white of the extrusion itself and of the background.

Power transmissions

When illustrating machinery the subject of transmission of power is often involved. This section describes just a few of the most common forms. Figure 6.26 shows the vee link belt drive which is a system of separate link units that can be assembled on site in a factory or workshop to suit any motor driven installation. Both light and heavy stipple have been used together with cross hatch to show the cross-section of the detail at the top right; otherwise the illustration relies completely on line.

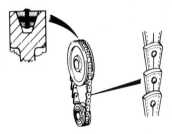

Figure 6.26. Vee-link belt drive with section detail top left

The toothed form of belt drive is designed not to slip, it is quiet in use and ideal for timing belts in today's automobiles. Figure 6.27 uses both line and stipple shading to clarify details of such a belt and the cross-section in situ on its pulley.

Next to gears, roller chains are the most common form of drive. The type we are all familiar with is the bicycle chain. Industrial developments of this simple drive system are shown in

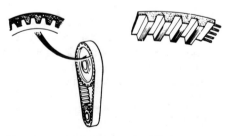

Figure 6.27. Toothed belt drive with tooth detail top right

Figure 6.28 and include single, duplex and triplex systems, for increased power transmitting capacity. They are always 'fiddly' things to illustrate and are best illustrated in line only, as here, unless large enough to have shading applied to each roller to help indicate its cylindrical form. Figure 6.29 indicates diagrammatically the gear tooth involute form. It should be

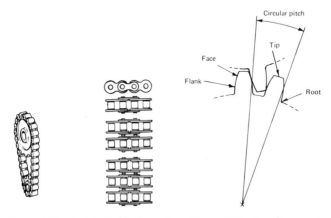

Figure 6.28. Roller chain drive of single, duplex or triple systems

Figure 6.29. The shape and main parts of gear teeth

noted that, contrary to the belief of many student technical illustrators, gear teeth are cut with a curved working face and are not flat faced at all. The form of the gear teeth is shown clearly in three dimensions in the next chapter (Figure 7.5 – right angle gearbox).

Gears

Gears are widely used to transmit parallel power in addition to various forms of non-parallel power. In their simplest form as straight toothed gears they are referred to as spur gears (Figure 6.30) and are widely used on machinery requiring transmission of a fixed ratio.

The helical gears shown in Figure 6.31 have similar applications but are much tougher and give a smoother operation than the equivalent spur gear transmisions.

Bevel gears have straight teeth, as do the spur gears, but are bevelled and are used to transmit power through 90° as shown in Figure 6.31. Obviously the gears having straight teeth are much easier to illustrate than those having curves, but all forms of gear trains present problems for the illustrator. Whilst it is not

Figure 6.30. A spur gear drive

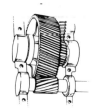

Figure 6.31. Part of a helical gear driven gearbox

Figure 6.32. Left, conical friction discs, and right, bevel gears with straight teeth

imperative to illustrate exactly the correct number of teeth on a gear wheel having a large number of teeth, the number must look correct. The best method, therefore, is to start from the point where the gears are meshing, and in most cases any discrepancy will be hidden behind some other piece of equipment. The meshing points can be seen on the illustrations of the spur gears and of the bevel gears (Figure 6.31 and 6.32 respectively). If however the gears are larger and their teeth less numerous the illustrator must be more precise in constructing the illustration. He or she has wider areas of the object on which to portray its shape and form, consequently more attention can be given to detailed shading resulting in better imagery! Observe the shading and precision of the illustration in Figure 6.33,

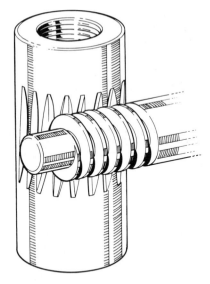

Figure 6.33. More shading can be applied to greater surface areas, resulting in better imagery

bearing in mind that if this gear combination was only a part of a larger unit then less attention could be directed to it.

Portraying translucency

The construction and illustration of gears demands great precision of the illustrator in the ability to project and produce exact detail. The ability to be a little more 'loose' and creative perhaps is required when confounded with translucent surfaces of glass, plastic or perspex.

Let us consider first a flat-surfaced clear vision guard, as used extensively in modern industry. Its object in Figure 6.34 is to protect the lathe operator from chips of flying swarf while affording the best possible view of the machining operation being carried out. Certain reflections are apparent on its surface, therefore only partial clarity through the surface is available. The reflection lines are best drawn as diagonals, in clusters from edge to edge.

The milling machine guard serves much the same purpose but it curves to enclose the circular milling cutters and the work being milled (Figure 6.35). The shading lines on the clear panels therefore must describe the curved surface whilst allowing a recognisable view of the operation being carried out.

Figure 6.34. Irregularly spaced angled line shading used to describe flat perspex, glass or transluscent surfaces

Figure 6.35. Irregularly spaced and broken parallel-to-axis lines describe the curved translucent surface

Figure 6.36. Shading the translucent handle of an electrician's screwdriver using an irregular line technique

We are all familiar with the screwdriver that has a clear plastic yellow handle (Figure 6.36), but have you ever tried to illustrate it convincingly? It has parallel grips on the handle which reflect both highlights and shadow onto the grip beneath. Not only is precision required of the illustrator here, but also the ability to observe light and shade in very close proximities; and, having only a pen making black markings on a white surface, it takes much practice to achieve a good result.

Electrics

In most cases the machinery that we technical illustrators portray is connected to or related to a source of power, that power being electricity. As illustrators of technical subjects, that of electrics in general is an area where our knowledge and ability must, through necessity, be strong. We must understand certain basic principles and how we can best communicate them to the layman. The electric plug shown at Figure 6.37 communicates certain basic principles. It is immediately recognisable with clean lines and good proportions. One detail which makes it seem

Figure 6.37. Electric plug, cable and wires

especially life-like, and which makes people look is the word 'FUSED'. Insulation wires have minor longitudinal shading, but otherwise the illustration relies on clean precise line work. Figure 6.38, showing three-core sheathed cable, is an example of the most neat and precise type of illustration. If colour or half tone cannot be used then the artist must rely entirely on the line and screens that are available. Here we have an example of just that, a clean, clear and unpretentious illustration that communicates well. See how different colours have been shown by the deft use of tones and line.

The sheathed cable at Figure 6.39 shows a different approach to the subject. Here the illustrator has indicated the coldness of the subject matter. The copper strands stand out from their insulation housing, while their outer skin is guarded by a copper

Figure 6.38. PVC sheathed cable

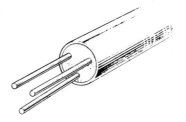

Figure 6.39. Copper sheathed cable

sheath. The shading tn the copper strands and the sheath communicates a smooth hard surface, while stipple has been used to show the insulation (magnesium oxide). Compare this treatment with that of Figure 6.38 which looks pliable and soft, which of course it is!

Welds

Another subject illustrators have to portray frequently is the welded joint. In most cases we are talking about heavy engineering subjects, and therefore the problem of showing vertical slabs of metal joined to horizontal baseplates by metal which has to look as if it is still molten. Uneven curves are used, hand drawn and irregular, after first pencilling out with a straight edge the weld run. Figure 6.40 is a good example that shows the welding

BASIC TECHNIQUES

Figure 6.40. Illustrating a weld

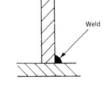

Figure 6.41. Weld shown orthographically

rod in action. Note the small amount of shading to draw attention to the weld. On the orthographic drawing the weld is shown as at Figure 6.41.

Ghosting

In Figure 6.42 the casting on the lathe faceplate has moved and the broken line or ghosted image shows where it was originally.

Figure 6.42. Ghosting

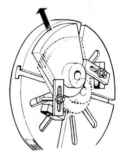

A different form of ghosting is used to give the impression of looking through part of an illustration when it can't be cut away, so that parts otherwise obscured can still be seen.

Enlarged detail

Very often the illustration contains more than one view of the subject when specific detail must accompany a general view. In Figure 6.43 the illustration would have to be enormous in order to show the shaping tool in detail if work and the tool post were

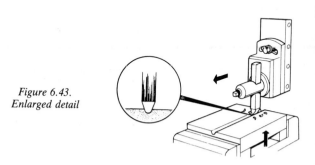

Figure 6.43. Enlarged detail

similarly enlarged. This alternative method means the illustration can be kept to a reasonable size while an orthographic enlargement gives the necessary tool tip detail. The milling cutter in Figure 6.44 is detailed in the same way. Note that in both cases stipple has been used to indicate the metal being machined in the orthographic detail enlargements.

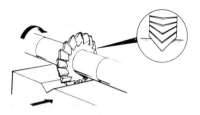

Figure 6.44. A milling machine arbor revolves whilst the metal being vee cut moves under the cutter. The cutter detail is enlarged orthographically

Springs

Springs are one of the most difficult of all the component parts that the student illustrator has to draw, but with practice it becomes just another item like a thread or a gear wheel. If the spring coils are drawn around another part e.g. a shaft passing through the central axis, then the coils furthest away inside the spring will be hidden from view, as in the case of Figure 6.45; the

Figure 6.45. The last couple of turns of the spring are hidden from view on the inside. If the ellipse angle were more shallow then even more would be hidden

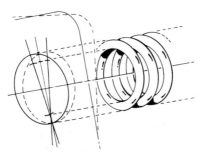

Figure 6.46. The inside and outside diameters together with half the pitch laid to left and half to right, then the commencement of the spring projection

number of coils hidden inside depends on the ellipse angle. If however, the spring is open and not in any way hidden (Figure 6.46), then your drawing must be completely accurate. The simplest construction method is as follows:
(1) Draw in the centre axis, outside and inside diameters.
(2) Mark off the length and pitch – the distance one coil is from the next coil – then draw in all the outside diameter ellipses.

(3) Draw in the inside diameter ellipses remembering that, since the coils are circular in section, the inside diameter ellipse must be the same distance away from the outside diameter ellipse all the way round. Usually you can accomplish this by changing the ellipse angle by five degrees, e.g. outer 45° – inner 40°.
(4) Finally link up the coils one to another on the back of the spring as shown in Figure 6.47.

Figure 6.47. Elaborate shading makes the spring really life-like

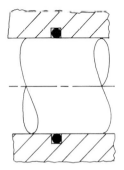

Figure 6.48. The 'O' ring is usually shown orthographically as two circular 'blobs'

The shading of springs, like that of threads, depends to a great extent on how large or small they are. If there is room then the spring can be made to look much more life-like (Figure 6.47).

'O' rings

After springs, 'O' rings are fairly easy to illustrate. They are manufactured from rubber and are used as a simple form of seal. A typical instance of their use is shown at Figure 6.48 where the 'O' ring is installed in an undercut of a housing in which a shaft is rotating. When illustrating the 'O' ring you must remember that it is circular in section and that if you drop back the ellipse angle by 5° from the outside to the inside diameter you can in most

Figure 6.49. The 'O' ring. When ellipse guides are being used make the outer ellipse 5° larger than the inner. This usually does the trick

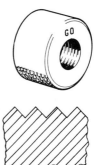

Figure 6.50. Cross-section through the diamond knurl and how it is shown in three dimensions

cases create the right illusion. A little stipple shading to indicate rubber completes the subject as shown in Figure 6.49.

Diamond knurl

A knurl is a roughening of the surface to give a better grip than a smooth surface would when handled. A cross-sectional and a three-dimensional view of a diamond knurl is shown in Figure 6.50. The whole of the outside surface of this ring guage is knurled but it is only necessary to show part of it when illustrating. Since the surface of the metal is pushed up into small 'diamonds', small 'vees' have been drawn on the bottom of each to add three-dimensional effect.

Worm drive clip

The clip shown in Figure 6.51 is frequently used for the joining of flexible hoses to metal pipes. The detail of the clip is shown in

Figure 6.51. Worm drive screw securing flexible hose

line only and the pipe has a little line treatment. The hose is stippled, affording the maximum contrast whilst retaining maximum clarity.

The circlip

The circlip is made of spring steel and is flat. It can be of external or internal type, Figure 6.52, and is often used for holding a ball race in position by 'springing' the circlip into a pre-machined undercut.

Figure 6.52. Internal and external circlips

7 Let's illustrate

There are two different methods of producing a technical illustration. One is to produce the work on line board or paper in pencil and then to complete it in ink, adding mechanical tints or tones (when necessary) at the same time as shading. The other method, and the method best suited to beginners, is to construct the work on tracing paper and then to 'push through' the image onto line board or paper and finally to ink. The second method affords the beginner the chance to make corrections at the pencil stage, and at a time when no damage is done to the material surface that is to carry the final inked image. For the beginner, then, I would suggest these stages:

(1)(a) Read and inwardly digest the orthographic information.
  (b) Analyse exactly what you are required to produce e.g. what are the most important items in the illustration.
  (c) Can you understand the size, shape and function of these items?
(2) Sketch free-hand itemising in each part of the illustration having previously decided the illustration is to be:
  (a) exploded
  (b) cut away
  (c) assembled (a three-dimensional external view only)
  (d) a combination of all three.
(3) Produce the grid suitable to give the best viewpoint for the subject and enabling the component parts to be viewed to the best advantage.
(4) Complete the pencilled construction illustration.
  (a) Insert any notes to yourself to remind you at the inking stage e.g.
    (i) translucent surface, research required.
    (ii) open up ellipse angle as axis recedes.
    (iii) extra perspective required.

(5) Push through (simplest form).
   (a) Take an ordinary blue pencil and scribble all over the BACK of your illustration so that a complete blue film covers the area.
   (b) Turn over and place the work on the board or paper on which the image is to be finished, and tape down securely, using masking tape NOT a clear adhesive tape having too strong an adhesive.
   (c) Take a 3H or 4H pencil and go over every line and comment on the drawing so that all the information is produced in blue on the inking board or paper beneath.
   (d) Carefully draw in centre lines and guide lines (accuracy is often lost at the pushing-through stage).
(6) Finish in ink.
   This is the most rewarding part of the whole sequence, the part you have constructed and worked for.
   (a) Work carefully and systematically.
   (b) Grease repels the ink so keep your fingers off the image area where possible.
   (c) Keep the work clean.
   (d) If your fingers perspire, cover the work with tracing paper and remove a small area to ink in, replacing the cover as you move on.
   (e) Refer to Chapter 6 for detail guidance (threads, springs, etc).
   (f) Arrange any annotation carefully and thoughtfully.

When smudges or errors are made at the ink stage, as they always are, then they must of course be rectified. An erasing shield can prove to be a good investment, as this will guard an area of your work while you erase the offending parts with an ink eraser. Alternatively, many students use a scalpel blade; but this tends to dig into the surface. I prefer the old-fashioned method of a razor blade. It can be held at a shallow angle, thereby affording the minimum damage to the surface by 'digging in'. Some of the older types are made of non-stainless steel, which is brittle and can be broken to give a sharp edge, while the newer types, when broken, cause a 'hook' to be formed at the break,

LET'S ILLUSTRATE

which 'digs in'. A sharp edge of course means that you can scrape ink away to an even finer degree.

The changing ellipse

Figure 7.1 shows the basic grid used previously. The three ellipses that go together to make up this grid are of correct angle within about 2°. I say within about 2° because the ellipse guides used in the grid's construction are of 5° changes e.g. 15°, 20°, 25°, 30° and so on (see Chapter 9, 'Tools of the trade' for full details).

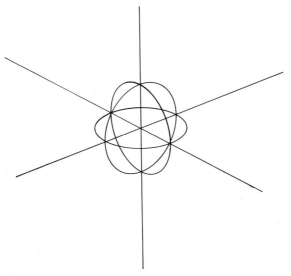

Figure 7.1. The basic grid

The horizontal ellipse is 25° but if the ellipse was moved higher up towards the eye level of the viewer it would become shallower – its minor axis would reduce. If however it was moved further away down its axis, it would become fatter – its minor axis would increase (Figure 7.2).

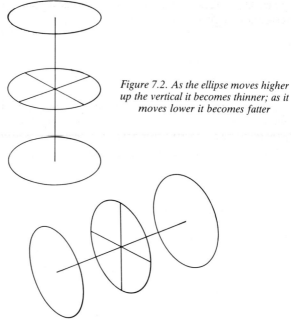

Figure 7.2. As the ellipse moves higher up the vertical it becomes thinner; as it moves lower it becomes fatter

Figure 7.3. Closer to the eye the ellipse narrows, further back it fattens

Figure 7.4. Move the ellipse forward and it slims; move it backwards and it fattens

The same changes take place along all three axes. Take the ellipses on the axis shown at Figure 7.3. The central ellipse on the grid was 35°, and as it moves forward it gets thinner; move it back along its axis again and it becomes fatter. The third ellipse in the grid was 45° and, as with the other two, move it forward and it slims down; move it backwards and it fattens up (Figure 7.4).

Obviously, whilst the ellipse gets closer to a circle as it goes further from the eye of the viewer, then at the same time it becomes perspectively smaller i.e. the major axis reduces in proportion to the minor axis. Ultimately, we can assume that there must be a time when, in the distance, all ellipses appear as circles in all three planes. But, for the subject level dealt with in this book, suffice it to say that these comments applied to Figures 7.1, 7.2, 7.3 and 7.4 must also apply to aid the production of convincingly acceptable small illustrations. The ellipse angle changes may cover more or less than 15° on any axis, depending on the size of the subject. Whilst you as the illustrator know that the ellipse angles must change, ordinary viewers don't necessarily understand. They only understand what they can see, so changes of ellipse angles are made where they are least noticed.

Right angle gearbox

Observe now Figure 7.5, the right angle gearbox. This unit is used to transmit power through 90°. Look at the vertical axis and count the number of ellipses along this central axis. While you work up from the bevel gears, pay attention to the ellipse angle changes as you climb closer to the eye level of the viewer. Notice the wide ellipse angle of the vertical axised bevel gear and the ball race above it compared with the ellipses of the parts higher up the axis and closer to the eye. See how shallow the angles of the nut and tab washer and the drive flange at the top of the illustration are when compared with the bevel gear at the base of the same vertical axis.

Observe now the front casing of the unit, the ellipse angle used for the cover, packing ring, screws and castle nut. Compare the ellipse angle of the component parts to the right of the illustrations, the drive flange, nut and tab washer. As before, the further from the eye then the fatter the ellipse becomes.

This illustration bears out therefore the points made in the section 'The changing ellipse': the ellipse angle does change, as you can see. So also do the perspective lines of the subject. The

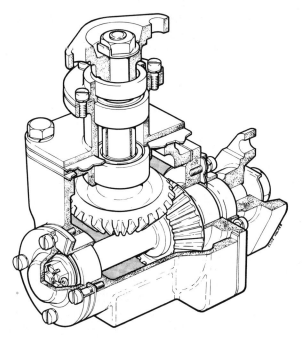

Figure 7.5. Right angle gearbox

right angle gearbox has a quarter section removed and all the cut-away surfaces are stippled for clarity. Compare the perspective lines to the right: the gearbox is only a small assembly but nevertheless a small amount of perspective has been added to satisfy the eye. To create interest and to clarify the various surface areas certain parts have been dragged out of the appropriate cutting plane. By doing this, the parts of the unit become much easier for the viewer to recognise, while depth has been given by the application of a dot screen (details at Chapter 9, 'Tools of the trade'). Otherwise shading is kept to a miminum. Compare the weights of line, outline with change of plane lines. If you could get your finger behind it then it is an outline or edge and must be 'heavied-up'. The small areas of shadow are really only 'hints of shadow' to help the three-dimensional effect. This is a good example of the standard of work to be expected by the completion of Technician Education Council Year III course.

Crane hook block

The City & Guilds of London Institute's part I examination paper of 1978 shows a crane hook block. Of the two main orthographic views the left (Figure 7.6) is completely sectioned down the centre line and reveals the simple workings of this unit. The small detail at section A–A shows the internal circlip securing the ball bearing, while the broken detail of the section view clarifies the final fixing, procedure of bearing pin, shouldered collar and socket screw.

To allow the hook freedom to turn it is mounted in a thrust ball race and secured by a castle nut which in turn is locked by a split pin. Thus all workings of the crane hook block are made plain. For further information e.g. two cross-sections at B–B and about the centre of the hook itself together with external details, we must refer to the right hand orthographic view (Figure 7.7).

The hook itself is probably cast but the shank is machined and threaded. The pulley also is probably cast, its inside diameter being machined to accept the two ball races and undercut to accept the two circlips. The pulley bearing pin could be turned from hexagonal bar while the shouldered collar head could be round or hexagonal.

To be able to produce an illustration of this subject all the information in the orthographic drawing must be assimilated and analysed by the illustrator who can then start work. One such piece of work is as shown at Figure 7.8 and is an example of a

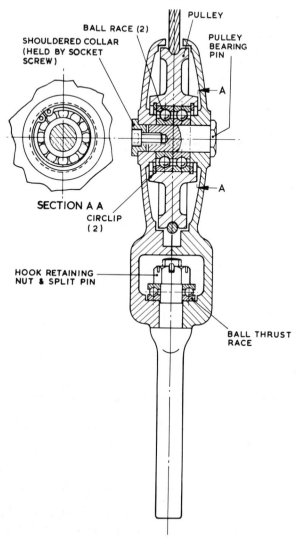

Figure 7.6. Exercise 1. Crane hook block – left orthographic view (courtesy City and Guilds of London Institute)

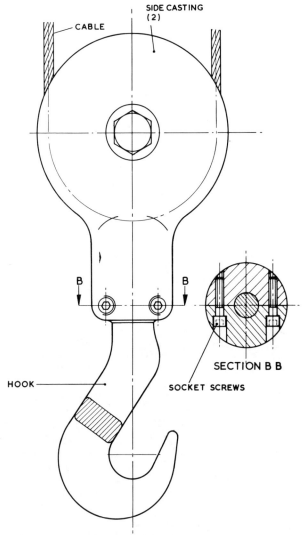

Figure 7.7. Exercise 1. Crane hook block – right orthographic view (courtesy City and Guilds of London Institute)

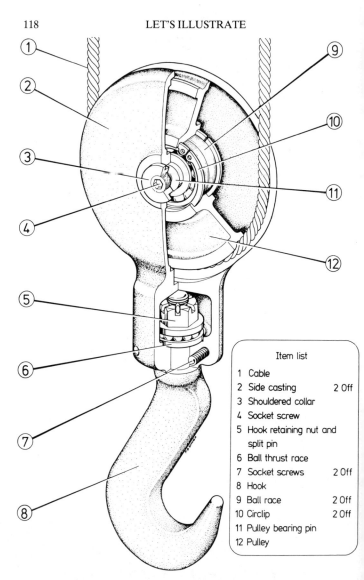

Figure 7.8. Crane hook block – cut-away

cut-away illustration produced from this orthographic information. The student illustrator has taken a quarter section of the front casting away to reveal the main working parts. The cut-away of the pulley and shoulder collar have been 'dragged out' of the section to add interest and to clarify the subject. Note that bearings, nuts and screws, are not sectioned and in this case note also the fact that the cut faces are left blank as a contrast to the perhaps too heavily stippled cast areas of the unit. Finally, observe the item list to which the numbers 1 to 12 apply. In this case the item list is stencilled with a number key; for reasons of clarity, this is preferable to placing named parts directly on the illustration. If the work contains a large number of parts, placing the names of each part directly onto the illustration creates confusion rather than clarifying the situation. It is *not* necessary to circle around each number (item blooming) but to use a tapered leader line of the weight shown here *is* a good practice. The line cannot be confused with any other line on the drawing and is therefore more easily read.

In 1974 the same examination had for its subject a small water pump, the details of which are shown at Figures 7.9 and 7.10. The orthographic information consists of the three main views (two of which are partially sectioned) and certain detail drawings. The pulley wheel (item 6) turns the cam (item 7) thereby transmitting the motion to the eccentric rod (item 16) which in turn, by the action of the plunger, creates a vacuum and lifts the ball on the valve (item 8). It is in fact a simple non-return valve system.

The illustration in Figure 7.11 shows one student's answer to the question. No stipple at all this time, although it is quite likely that the body would be cast; instead, line shading has been used to help describe the shape. It has also been used on the drive pulley, eccentric strap and gland nut, while the cut-away section is shown by the use of a tone or dot screen. Although the item numbers are arranged tidily around the illustration, the item list is too far away from the items. This does cause a little confusion e.g. one or two gland nuts?

Use the orthographic information for these two subjects, together with the illustrations as a guide, to produce your own

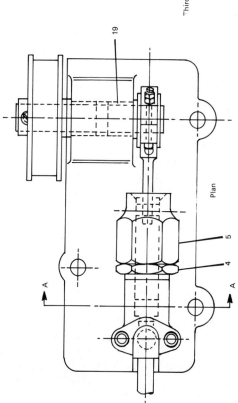

LET'S ILLUSTRATE

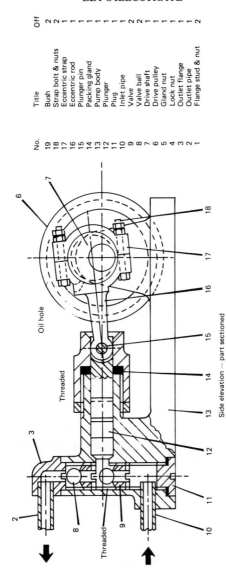

Figure 7.9. Exercise 2. Water pump (courtesy City and Guilds of London Institute)

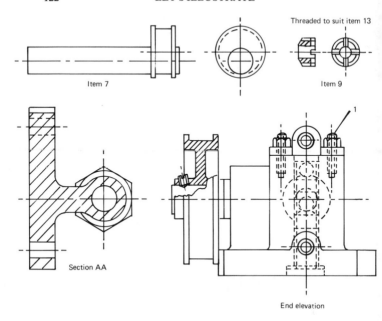

Figure 7.10. Exercise 2. Water pump (courtesy City and Guilds of London Institute)

illustrations using methods already described. When these two have been completed, work your way through the following subjects. Don't worry about speed but concentrate on quality. Keep all your work and you will surprise yourself at the improvements you make as you progress by reading the orthographic drawing first, producing a free-hand sketch and finally illustrating.

Since all these subjects have had to be reduced to comply with the page format for this book I would suggest that you scale up in order to produce illustrations of a reasonable size to work on. Some reduction in size will occur as you foreshorten on the grid so a suggested scale from the orthographic is three times up. You may prefer to scale up differently for the different subjects.

LET'S ILLUSTRATE

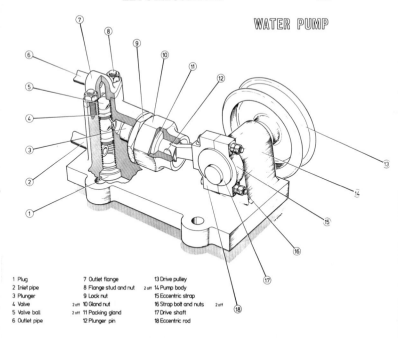

WATER PUMP

1 Plug	7 Outlet flange	13 Drive pulley
2 Inlet pipe	8 Flange stud and nut 2 off	14 Pump body
3 Plunger	9 Lock nut	15 Eccentric strap
4 Valve	2 off 10 Gland nut	16 Strap bolt and nuts 2 off
5 Valve ball	2 off 11 Packing gland	17 Drive shaft
6 Outlet pipe	12 Plunger pin	18 Eccentric rod

Figure 7.11

Final illustrating exercises – orthographic information

Complete the odd numbered titles as exploded illustrations and the even ones as cut-aways.

(1) Crane hook block	Figures 7.6	and 7.7
(2) Water pump	7.9	and 7.10
(3) Pipe vice	7.12	and 7.13
(4) Squeeze riveter cylinder	7.14	and 7.15
(5) Expansion joint	7.16	and 7.17
(6) Drain cock	7.18	and 7.19
(7) Seal puller	7.20	and 7.21
(8) Geneva mechanism	7.22	and 7.23
(9) Tank contents gauge	7.24	and 7.25

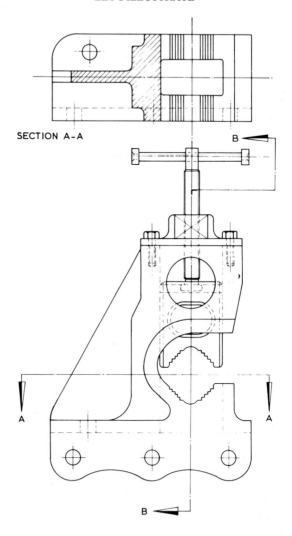

Figure 7.12. Exercise 3. Pipe vice (courtesy City and Guilds of London Institute)

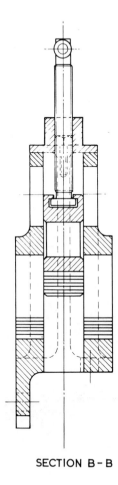

SECTION B-B

Figure 7.13. Exercise 3. Pipe vice (courtesy City and Guilds of London Institute)

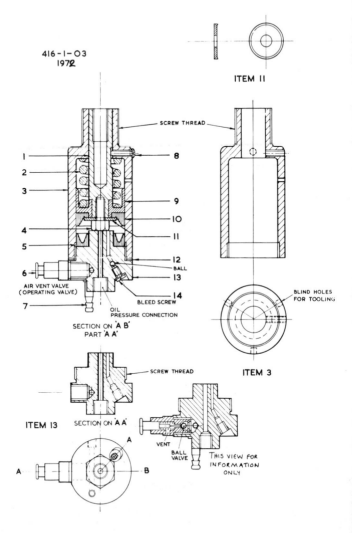

Figure 7.14. Exercise 4. Squeeze riveter cylinder (courtesy City and Guilds of London Institute)

LET'S ILLUSTRATE

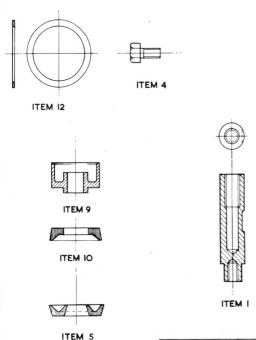

ITEM	DESCRIPTION	MATERIAL	No OFF
1	PISTON SHAFT	CAST IRON	1
2	SPRING	SPRING STEEL	1
3	CYLINDER	DURAL	1
4	BOLT	MILD STEEL	1
5	SEALING WASHER	RUBBER	1
6	VALVE ASSEMBLY	DURAL	1
7	CONNECTOR	BRASS	1
8	SCREW	MILD STEEL	1
9	PISTON	CAST IRON	1
10	SEALING WASHER	RUBBER	1
11	WASHER	MILD STEEL	1
12	WASHER	COPPER	2
13	CYLINDER HEAD	CAST IRON	1
14	BLEED SCREW & BOLT	MILD STEEL	1

Figure 7.15. Exercise 4. Squeeze riveter cylinder (courtesy City and Guilds of London Institute)

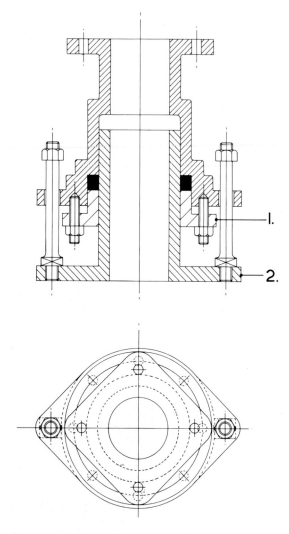

Figure 7.16. Exercise 5. Expansion joint (courtesy City and Guilds of London Institute)

Expansion Joint.
Scale 1:1.

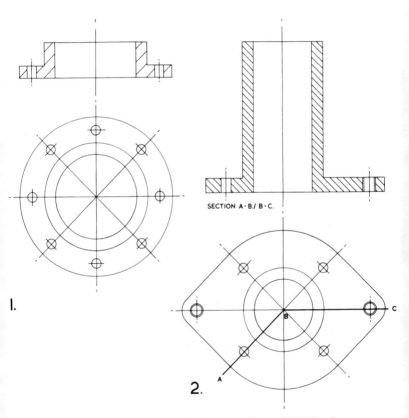

Figure 7.17. Exercise 5. Expansion joint (courtesy City and Guilds of London Institute)

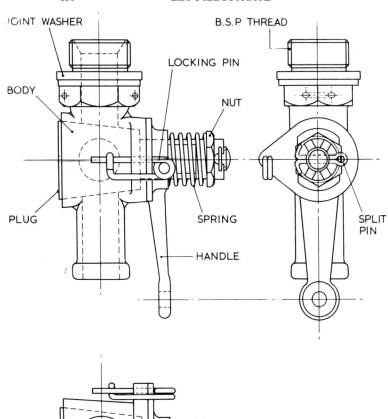

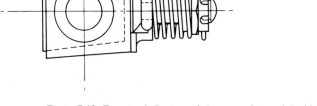

Figure 7.18. Exercise 6. Drain cock (courtesy City and Guilds of London Institute)

LET'S ILLUSTRATE

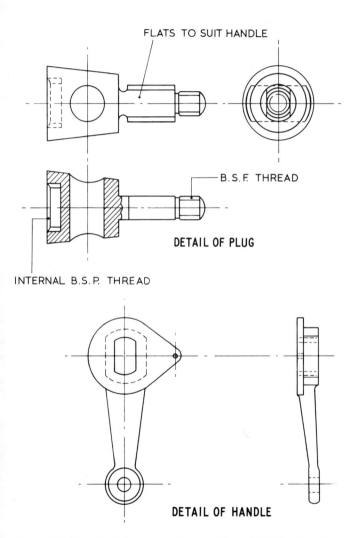

Figure 7.19. Exercise 6. Drain cock (courtesy City and Guilds of London Institute)

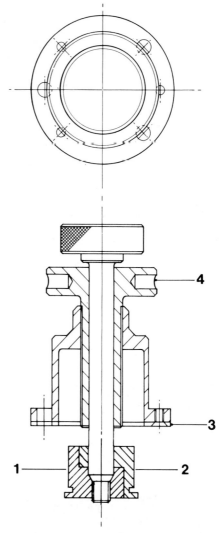

Figure 7.20. Exercise 7. Seal puller (courtesy City and Guilds of London Institute)

LET'S ILLUSTRATE

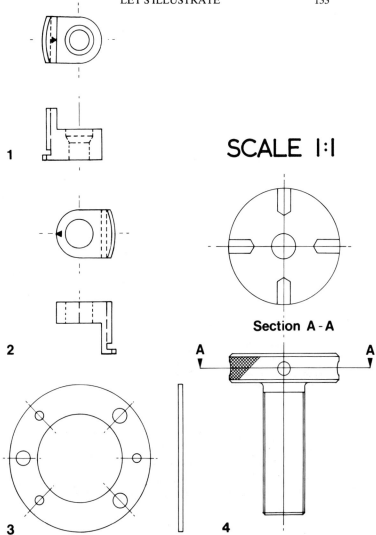

Figure 7.21. Exercise 7. Seal puller (courtesy City and Guilds Institute of London)

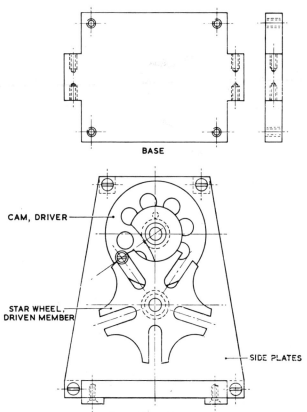

Figure 7.22. Exercise 8. Geneva mechanism (courtesy City and Guilds of London Institute)

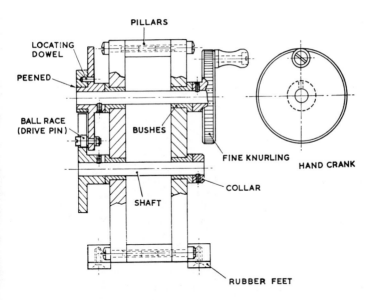

Figure 7.23. Exercise 8. Geneva mechanism (courtesy City and Guilds of London Institute)

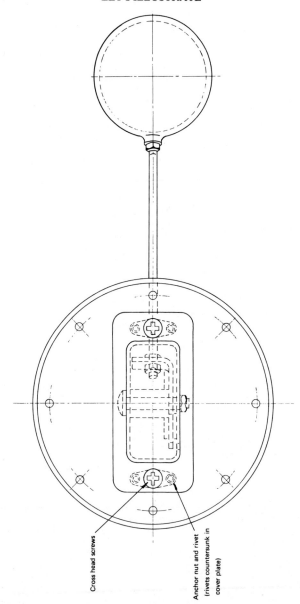

LET'S ILLUSTRATE

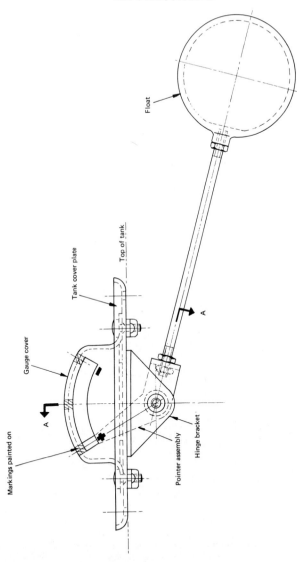

Figure 7.24. Exercise 9. Tank contents gauge (courtesy City and Guilds of London Institute)

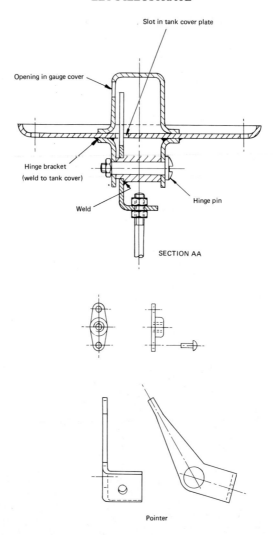

Figure 7.25. Exercise 9. Tank contents gauge (courtesy City and Guilds of London Institute)

8 Free-hand drawing and sketching

Some people are certainly more fortunate than others when it comes to this part of the subject. The lucky ones seem able immediately to draw an image just as it is seen, whether it be a landscape or a motor cycle! However, the vast majority are not so fortunate, which means, of course, that for the would-be technical illustrator the basic principles must be understood and a great deal of time spent on practice in order to master free-hand methods with which he can communicate technical information before commencing a constructed illustration. If the illustrator is instructed by the compiler or by the technical writer that a certain illustration must, show say, fifty-two items, then in order to locate those items and in order to work out the layout for his illustration, he must first of all produce a free-hand sketch. This would be for his own benefit but very often he will be working with authors or copywriters and deciding on the type and style of illustration required. Which viewpoint gives the best result, which one the most important features? Where to 'cut-away', where to 'explode'? Are additional flat diagrams necessary to show the function? The illustrator cannot start until these questions are answered, and his ability to sketch at this stage is imperative. Alternatively he may be sketching information and recording notes, written and photographic, when no orthographic information is available, so that at a *later date* illustrations can be produced by *another* artist. The ability to produce informative free-hand drawings, therefore, from the object, from memory, from orthographic and other sources is a must.

As a student you should keep a sketch pad and a couple of pencils handy. An F and an H grade are as good as any to start with, and you should allocate yourself some time to practise

140 FREE-HAND DRAWING AND SKETCHING

sketching, maybe during lunch-breaks, maybe a couple of evenings a week. Sketch the outside of your house and the view down the street, looking carefully at the perspective lines. Sketch in the car park and in the shopping centre; sketch buses, trucks and trees. After working on a variety of subjects then narrow down the field to mainly mechanical subjects.

When commencing a course of study on technical illustration the majority of students forget that certain basic rules apply to both constructed illustrations and free-hand drawings. They are not two entirely different subjects (as are welding and domestic science for instance). Earlier in this book we have dealt with the ellipse, how it is constructed and how it is used, but many students seem to forget that when free-handing a simple bracket such as the one dealt with in Chapter 4 (Figure 4.25), the major and minor axis lines are still at right angles and should still be drawn in as at Figure 8.1. Having decided on the major axis this

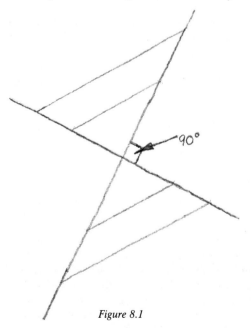

Figure 8.1

FREE-HAND DRAWING AND SKETCHING 141

can be marked off, half each side of the centre; and lines can be sketched in linking major to minor dimensions as a guide for the next stage as at Figure 8.2, the completion of the ellipses.

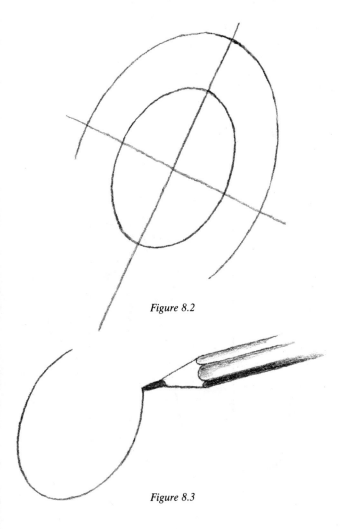

Figure 8.2

Figure 8.3

Always hold the pencil lightly for free-handing, a little further from the point than you would for writing so that you can 'see where you are going'; if your hand is in the way you can't. Short strokes are used for free-handing, one joined to the other; never attempt to complete an ellipse for instance as in Figure 8.3 (in one fell swoop). When completing a straight line try using 'overlapping' strokes. For long lines keep the arm stiff rather than using a finger- or wrist-movement. Concentrate the eye meanwhile on the point at which the line is aimed and not on the pencil point, and draw away from the body for straight lines.

Proportion is another important point that beginners pretend isn't there at all. Length, width and height again are just as important as in constructed drawing. Look at the proportions of

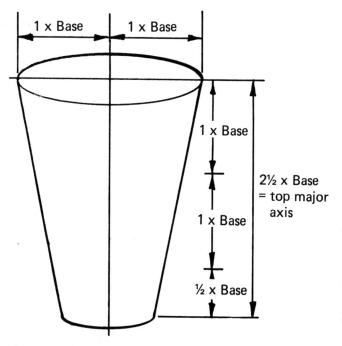

Figure 8.4. Correct proportions by eye

the object in Figure 8.4, and note the major axis of the top in relation to that of the base. It must be apparent to the viewer that the top is twice the base and that the object is two and a half bases high, just as when drawing a bolt head of hexagon shape. If it looks to the viewer that it has five sides then you as the illustrator are not communicating properly. When free-hand drawing, you are portraying what you are looking at; therefore stop and think of what you are doing. Call upon the knowledge you have already gained: you know of plan and elevation of the subject; you know of proportion; you know how to construct. Put it all together and communicate to the viewer of your work what he or she doesn't know.

When free-handing, never start working in detail at one end of the drawing and then working to the other parts from there. If you do this there is a good chance that by the time you get to the end it will be out of proportion. Instead block in the main features and proportions of the object first, then when you are happy with it you can start to get down to the details, gradually putting it together one area in relation to the other. In Figure 8.5 we see the outline of the figure completed, and the height of the figure is indicated in the number of heads high, in this case eight and a half. In Figure 8.6 the sketch is completed on top of the preliminary 'working out' of Figure 8.5 so we know that the girl hasn't grown too tall, or too short.

Finally, let us go back to the simple bracket we discussed earlier and look at a system of blocking in an overall impression of its shape, size and proportion. Figure 8.7 gives an indication of 'blocking in' length, width to height. How many 'holes' high is the height of the web? How many holes wide? What percentage of the total height is the plate thickness? Look and compare.

Figure 8.8 shows a final development of our first impression. We have evolved the heights, widths, lengths and proportions. The two holes in the front of the base plate are lined up as a proportional distance from the front edge, their centres and axes assessed, and then they are completed. The top radius of the web and the blend radius are at last completed, as is our lesson in free-hand drawing and sketching of technical matter – save for these comments.

Figure 8.5. Initial 'blocking in'

FREE-HAND DRAWING AND SKETCHING

Figure 8.6. Finishing off after 'blocking in'

Although technical illustrators often give the impression that their subject is worlds apart from any other branch of the visual arts, this is not the case. So go to galleries and exhibitions and observe the work of other artists, designers, engineers and architects and if you have the chance take a look at the drawings and paintings of Leonardo da Vinci, of Picasso, of Rembrandt and of any of the other accepted masters; then you will really begin to understand what is meant by drawings and sketching.

In the meantime the following list of free-hand drawing subjects should give you plenty of practice. Aim to complete them on A3 size paper, drawing in line only to begin with so that your work has to be good enough to be recognised wihout shading. Later, when you are more experienced, you can add shading and tone.

Don't be mean about the drawings; you are not designing postage stamps. Fill the A3 size paper and give the viewer something to look at.

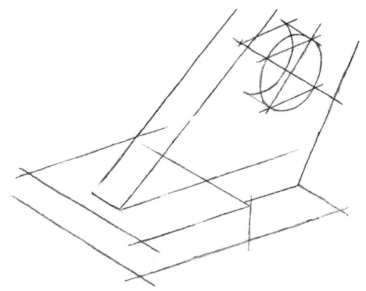

Figure 8.7. Initial 'blocking in'

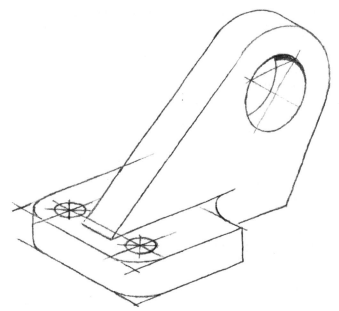

Figure 8.8. Finishing off after 'blocking in'

Free-hand drawing from memory (pencils and erasers only to be used)

(1) a woman pushing a pram.
(2) a horse chestnut tree, showing details of leaf structure, seed and seed coat.
(3) a covered bus shelter, with seat and waste paper receptacle.
(4) a stationary motor cycle on its stand.
(5) a musical instrument, such as a violin.
(6) an aircraft on a runway.
(7) greenhouse interior with plants 'in situ'.
(8) steam engine, e.g. locomotive, traction engine, showman's engine.

(9) children in a play area with at least three pieces of mechanical amusement, e.g. slide, roundabout, rocking horse, climbing frame.
(10) corporation refuse lorry with personnel.
(11) typical roadworks including workmen, equipment and surrounding buildings or landscape.
(12) the view from the top of a staircase in a department store looking down and including the lower ground floor.
(13) a woman using a lawn mower.
(14) a record player turntable and pick-up arm.
(15) the interior of a supermarket, showing display shelves, cash desks etc.
(16) a small printing press.
(17) an oak tree, giving details of the leaf, flower and fruit.
(18) a school coach.
(19) a milk float.
(20) a filling station forecourt, complete with petrol pumps.
(21) a rowing boat.
(22) a bus station (not a shelter).
(23) interior of a 'village store'.
(24) domestic fuel delivery tanker.
(25) a car ferry unloading.
(26) a modern kitchen.
(27) a diesel locomotive.
(28) a railway station booking hall.
(29) a hotel lounge.
(30) a suburban station scene.
(31) a fire engine and personnel.
(32) an ice-cream van and children.
(33) the scene at an airport lounge.
(34) a sailing dinghy.
(35) a modern car towing a caravan.
(36) a motorway intersection.
(37) an angle-poise type lamp.
(38) a ticket collector and immediate surroundings at a railway station.
(39) a small hardware shop with goods on display outside.
(40) the interior of a nursery greenhouse.

FREE-HAND DRAWING AND SKETCHING

(41) a person changing the rear wheel of a car.
(42) the interior of a school gymnasium.
(43) the supermarket checkout.
(44) a kitchen interior.
(45) three-quarter view of a motor car with drivers' door and boot open.
(46) a pipe vice.
(47) a child's tricycle.
(48) a child's pedal car.
(49) a typewriter.
(50) inside of a country inn.
(51) at the docks.
(52) the child's playground.
(53) inside the post office.
(54) a telephone kiosk.
(55) a domestic food mixer.
(56) a farm tractor and driver.
(57) a rotavator.
(58) an ambulance with rear doors open.
(59) a scene at the locks on an inland waterway.

9 Tools of the trade

Certain instruments are essential in our business; others, while not being essential, certainly make life easier. Twenty years ago many older illustrators still used a mapping pen and brushes for the purpose of applying ink. Since then the amount of equipment available has become enormous, and this section can only deal with a very small amount of that which is on the market. For the beginner the best thing to do is build up the tools of the trade a few at a time, as cost is a not unimportant factor when deciding what to buy.

Paper and board

Never try to ink on cartridge paper, or for that matter, any other type of paper or board not specifically made for inking. The paper or board used has a hard kaolin-coated surface and, whilst being absorbent to a certain extent, the ink does not 'feather' at the edges like ink on blotting paper. Also, when ink is scraped off, the surface can be pressed down again by using a hard smooth-surfaced object over a piece of thin paper, after which the ink can be re-applied. It will be found that different manufacturers charge a variety of prices for essentially the same material; a good idea is to try a variety and find which best suits your pen and your pocket.

For those not familiar with the international paper sizes a list is shown opposite. It should be noted that each size is half that of the preceding one.

TOOLS OF THE TRADE   151

'A' sizes	Approx. equivalent inches
A0 841 × 1189 mm	33.11 × 46.81
A1 594 × 841 mm	23.39 × 33.11
A2 420 × 594 mm	16.54 × 23.39
A3 207 × 420 mm	11.69 × 16.54
A4 210 × 297 mm	8.27 × 11.69
A5 148 × 210 mm	5.83 × 8.27

Don't use drawing pins on the drawing board; use masking tape as it has less adhesive than the clear adhesive types of tape and therefore causes less damage to artwork on removal. Use clear adhesive tape when a permanent fixing is required, such as taping covers on finished artwork.

Pencils

The humble wooden shelled graphite pencil (Figure 9.1) has to a great extent been superceded in recent years by the mechanical pencil (Figure 9.2). When a wooden pencil gets blunt it has to be

Figure 9.1. The wooden pencil

sharpened, first with a knife and then on a sandpaper block to arrive at the desired graphite tip. The modern mechanical pencils however require only a touch on the button with the thumb to supply lead to the drawing tip, and as they are available in a variety of lead sizes, including 0.3 mm, 0.5 mm, 0.7 mm and 0.9 mm, sharpening is unnecessary (Figure 9.2).

Figure 9.2. The fine lead holder or 'mechanical pencil' requires only a touch on the button to supply the lead, see left

The clutch pencil or lead holder (Figure 9.3) is really a pencil somewhere between the wooden pencil and the mechanical pencil. It is designed to use a larger diameter lead than the mechanical pencil, in some cases from 1.9 mm to 3.15 mm. Since

Figure 9.3. The clutch pencil

the lead is this size, it still needs sharpening and a sandpaper block is advisable. As with all these pencils a wide range of leads are available, although you may find that one make of 3H is not as hard as another 3H for instance.

Erasers

The most useful eraser I have found is a dual-purpose one of vinyl. One third of it is hard for ink and the other two thirds is soft for pencil (Figure 9.4). Also on the market are eraser

Figure 9.4. Dual-purpose eraser, one end for ink, the other for pencil

holders into which are fitted the eraser cores for use on either pencil or ink. The core tip can be trimmed to a point to facilitate correction of work in a small area (Figure 9.5).

Figure 9.5. Eraser holder with erasing tip secured by the collar

Erasing shield

Figure 9.6 shows an erasing shield which is in fact a thin piece of stainless steel that has a number of small shapes pressed out of its

Figure 9.6. The erasing shield

surface. It is used as a guard for your work as you erase through the space, thereby inflicting the minimum of damage to the surrounding area.

Scalpel

Otherwise known as a surgical blade and handle, the scalpel (Figure 9.7) is used for a hundred and one jobs, from cutting board to trimming down mechanical tints on the illustration. A selection of interchangeable blades are available.

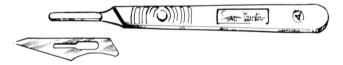

Figure 9.7. Scalpel handle with blade

Dividers

The dividers are used to transfer dimensions from the orthographic drawing to the illustration, or from the rule to the

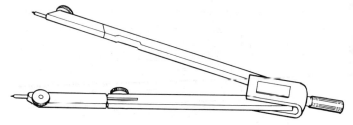

Figure 9.8. Geared head dividers

illustration, and consequently they are equipped with two fine needle points. The type shown in Figure 9.8 are of the geared head type.

Proportional dividers

Proportional dividers' prime use for the illustrator is in scaling up on dimensions. By making the appropriate setting on the scale, a dimension can be taken with one end of the divider, and the other end enlarges the dimension 3:1, 4:1 or whatever the scaling is set for (Figure 9.9).

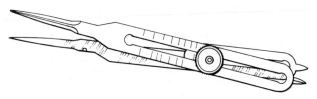

Figure 9.9. Proportional dividers of traditional design

Brushes

From time to time you will find a few brushes useful, mainly for whiting-out highlights or putting in a white 'shadow' beneath a leader line. You won't use a brush very often since the technical pens cover such a range of line thickness, but I always find a

Figure 9.10. Sable haired brush

sable haired '00' size very handy to have amongst my kit (Figure 9.10) as it has a fine point, is strong and springier than other hair brushes.

Compass or springbow

Which type to choose really depends on the individual. Most types have interchangeable pencil, pen and divider as in Figure 9.11. A recent development is a compass that fits the technical pens, thereby eliminating the need for the ink ruling pen.

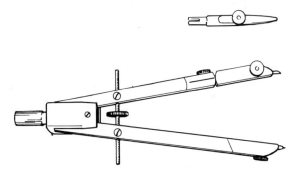

Figure 9.11. Springbow type compass with pen or pencil attachments

Ruler

The perspex rules having a bevel on both edges are the most useful, being divided into 64ths of inches and mm/cm. Four lengths are available, 12, 15, 18 and 25 inches (Figure 9.12).

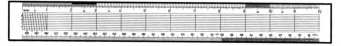

Figure 9.12. The 12 inch perspex rule

French curves

Very valuable to the illustrator is a good set of curves manufactured from non-chip pliable plastic. Some of the cheaper sets only contain three curves (Figure 9.13) but the larger the set you

Figure 9.13. Three typical french curves

can get, the more useful you will find them. A set comprising of some french curves and some slow curves will be found to be the most useful.

Adjustable square

Some adjustable squares are made from coloured and some from clear plastics, all having a bevelled edge. Normally, a white opaque plastic octant is used with scales being divided into half

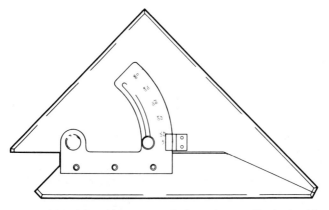

Figure 9.14. Adjustable square made from clear acrylic plastic with bevelled edge

degrees. Fittings can be steel, brass or aluminium (Figure 9.14). The 10 inch (250 mm) size is about the smallest useful size for our purpose. Don't bother with set-squares as they are non-adjustable, so are of little use as most of our work, unlike draughting, is rarely at standard angles.

Ellipse guide templates

These are probably the most expensive item of equipment you need and are normally produced as a ten-piece set in two

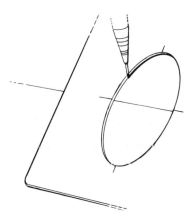

Figure 9.15. Pencilling with the ellipse guide

sections. The older designs have a major axis from about ⅛ inch to 2 inch from 15° to 60°. A larger set goes up from 2⅛ inch to 4 inch major axis, again in 15° to 60° all in 5° steps. A newer version on the market has the same ellipse angles 15° to 60° in steps of 5° but is dimensioned in metric and rises from 8 mm to 60 mm major axis. They are about 0.6 mm thick in a tinted plastic which allows

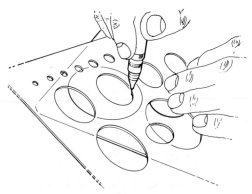

Figure 9.16. Inking with the ellipse guide

good visibility through the template to construction lines beneath (Figure 9.15). The major and minor axis being constructed in the manner previously described, and having proved that the ellipse angle is correct according to the constructed grid, the template is then laid down so that major and minor axis lines on the template line up with those constructed and the ellipse drawn in. When inking, however, it is best to raise the template off the working surface to avoid ink running underneath and smudging the work. This can be done by placing a larger degree ellipse under the one being used as in Figure 9.16.

Technical pens

The modern drawing pens, or technical pens as they are otherwise known, consist of four basic parts:

(a) the cap
(b) the handle
(c) the reservoir
(d) the nib

Item (d) is really the working part and is essentially a tube with a wire down the centre which helps the ink on its way from reservoir to working surface. This simple idea has been developed so far that you can now produce lines of uniform density

Figure 9.17. The working end of the technical pen with an indication of the line thickness it produces. Pens are colour coded in addition to having the size in mm on the nib

from a thickness of 0.1 mm right up to 1.2 mm thickness using ten different nibs for one particular make of pen (Figure 9.17). Different ranges of different makes all having different uses give an endless number of line thickness.

Inks

Although we occasionally use coloured inks, the majority of our work is produced in black on a white ground. Always make sure, therefore, that the correct black drawing ink for line paper or board is used, as the wrong ink can clog the pen.

Stencils

Like the drawing pens, the stencil pens and the stencils themselves have been developed tremendously in recent years, and

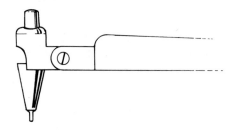

Figure 9.18. The modern stencil pen

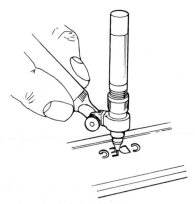

Figure 9.19. A drawing pen designed for use also as a stencil pen

they have become much less crude instruments than their forerunners. Many are moulded in tough plastic with steel tubes and push button centre wires as in Figure 9.18. Figure 9.19 shows a drawing pen which is specially designed to be used for stencilling as well. Today's stencils cover a very wide range of typefaces, some of them being condensed and some extended. The stencil is used extensively on artwork for technical publications as it is easy to use, and once purchased only the cost of the ink is extra. Like ellipse templates they will last a long time if taken care of.

Dry transfer lettering

A very comprehensive range of typefaces is now available from a number of manufacturers all competing with each other for a slice of this lucrative market; consequently a mind-boggling number of traditional as well as modern styles are available.

To use dry transfer lettering simply position the letter (Figure 9.20), rub over the front of the plastic sheet using a burnisher, ball point pen or pencil (Figure 9.21) and remove the sheet

Figure 9.20. Positioning the sheet of dry transfer lettering

carefully. When the whole word is complete, take the backing sheet that is supplied with the type sheet, place it over the word you have set and bone down i.e. rub all over it again. When corrections have to be made, remove the offending characters

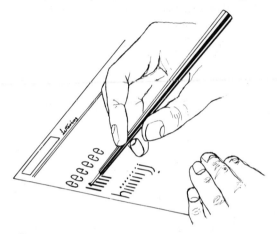

Figure 9.21. Rubbing over the required letters with the burnisher after positioning

with a low tack adhesive tape e.g. masking tape. Do this before the final boning down.

Shading film

Often referred to as mechanical tints or dot and line screens, a tremendous selection is available to the illustrator, but for black illustrations on white ground the most useful are the dot screens from superfine through to almost pure black. Figure 9.22 shows some of these available. The application is as follows:
(1) Score a section of film slightly larger than the area required and peel from the backing sheet.
(2) Position the film on the artwork.
(3) Trim excess film and smooth from the centre to exclude air bubbles on the area where the film has been placed.

Figure 9.22. Just a few of the dot screens available

Index

Adjustable square, 157
Annotation, 110

Bearings, 91
Bolts, 88, 89
Brushes, 155

Circlips, 108
Circular objects, 92
Compasses, 155
Crane hook block, 115–118

Dividers
  geared head, 154
  proportional, 154
Draughting, 19, 20

Electrical components, 101, 102
Elevations
  end, 10–15, 49
  front, 10–15, 49
  plan, 10–15, 49
Ellipse, 30–35
  changing shape, 111–113
  construction, 32–34
  guides, 157–159
  isometric, 40
Enlarged detail, 104
Erasers, 152
Erasing shield, 153

First angle projection, 12–19
'Flatties', 1
Foreshortening method, 50–61
  isometric projection, 38, 39

Free-hand drawing, 139
  blocking-in, 143–146
  proportion, 141, 142
  subjects to draw, 147–149
French curves, 156

Gearbox, right-angle, 113, 114
Gears, 97–99, 113
Ghosting, 103

Ink, 160
Intersections, 67–69
Isometric
  ellipse, 40
  foreshortening, 38–39
  projection, 36–39, 41
  square, 40
  view, 40

Knurl, 107

Lettering, dry transfer, 161, 162
Line drawing, 2
  shading, 93

Nuts, 88, 89

'O' ring, 106
Orthographic drawing, 8
  end elevation, 10, 11
  first angle projection, 12, 13
  front elevation, 10, 11
  plan, 10, 11
  third angle projection, 12, 13

INDEX

Orthographic exercises
   questions, 16–18
   answers, 26, 27

Paper, 150
   sizes, 151
Pencils, 151
Pens, 159
Perspective, 75
   single point, 76, 77
   two point, 77, 78
   three point, 78, 79
Perspective grid, 75
Projection,
   axonometric, 40, 41
   dimetric, 41
   oblique, 40
Proportional grid, 54–56
   foreshortening, 55
   illustration, 61–63
   use of, 59, 61–67

Radii, 84–86
Right-angle gearbox, 113, 14
Rulers, 156

Scalpel, 153
Screws, 88, 89
Screwdrivers, 98

Shading, 4, 93
   line, 93
   stipple, 94, 108
   techniques, 93–95, 99
Shading screens, 162, 163
Sketching, *see* Free-handing
Springbow, 155
Springs, 105, 106
Squares, adjustable, 157
Stages of illustration, 109, 110
Stencils, 160, 161

Tables, conversion, 20, 21
Technical illustrations
   cut-away, 2
   exploded, 2, 3
   ghosted, 103
Technical pens, 159
Third angle projection, 12–19, 33, 120
Threads, 86, 87
Trammel, 32–34
Translucency, 4, 99, 100

Vanishing point, 76–80

Washers, 90
Water pump, 120–123
Welds, 102
Worm drive clip, 107